# Wet Prairie

The Nature | History | Society series is devoted to the publication of high-quality scholarship in environmental history and allied fields. Its broad compass is signalled by its title: nature because it takes the natural world seriously; history because it aims to foster work that has temporal depth; and society because its essential concern is with the interface between nature and society, broadly conceived. The series is avowedly interdisciplinary and is open to the work of anthropologists, ecologists, historians, geographers, literary scholars, political scientists, sociologists, and others whose interests resonate with its mandate. It offers a timely outlet for lively, innovative, and well-written work on the interaction of people and nature through time in North America.

General Editor: Graeme Wynn, University of British Columbia

Claire Elizabeth Campbell, *Shaped by the West Wind: Nature and History in Georgian Bay*

Tina Loo, *States of Nature: Conserving Canada's Wildlife in the Twentieth Century*

Jamie Benidickson, *The Culture of Flushing: A Social and Legal History of Sewage*

William J. Turkel, *The Archive of Place: Unearthing the Pasts of the Chilcotin Plateau*

John Sandlos, *Hunters at the Margin: Native People and Wildlife Conservation in the Northwest Territories*

James Murton, *Creating a Modern Countryside: Liberalism and Land Resettlement in British Columbia*

Greg Gillespie, *Hunting for Empire: Narratives of Sport in Rupert's Land, 1840-70*

Stephen J. Pyne, *Awful Splendour: A Fire History of Canada*

Hans M. Carlson, *Home Is the Hunter: The James Bay Cree and Their Land*

Liza Piper, *The Industrial Transformation of Subarctic Canada*

Sharon Wall, *The Nurture of Nature: Childhood, Antimodernism, and Ontario Summer Camps, 1920-55*

Joy Parr, *Sensing Changes: Technologies, Environments, and the Everyday, 1953-2003*

Jamie Linton, *What Is Water? The History of a Modern Abstraction*

Dean Bavington, *Managed Annihilation: An Unnatural History of the Newfoundland Cod Collapse*

J. Keri Cronin, *Manufacturing National Park Nature: Photography, Ecology, and the Wilderness Industry of Jasper*

NATURE | HISTORY | SOCIETY

# Wet Prairie

## People, Land, and Water in Agricultural Manitoba

SHANNON STUNDEN BOWER

FOREWORD BY GRAEME WYNN

UBC Press • Vancouver • Toronto

21 20 19 18 17 16 15 14 13 12 11     5 4 3 2 1

Printed in Canada on FSC-certified ancient-forest-free paper
(100% post-consumer recycled) that is processed chlorine- and acid-free.

---

**Library and Archives Canada Cataloguing in Publication**

Stunden Bower, Shannon,
    Wet Prairie : people, land, and water in agricultural Manitoba /
Shannon Stunden Bower.

(Nature, history, and society series, 1713-6687)
Includes bibliographical references and index.
ISBN 978-0-7748-1852-0

    1. Agriculture – Manitoba – History. 2. Drainage – Manitoba – History.
3. Watershed management – Manitoba – History. 4. Agriculture and state –
Manitoba. 5. Agriculture – Social aspects – Manitoba. I. Title. II. Series: Nature,
history, society

HD1683.C3S78 2011          333.76′097127          C2011-900886-6

Canadä

UBC Press gratefully acknowledges the financial support for our publishing program of the Government of Canada (through the Canada Book Fund), the Canada Council for the Arts, and the British Columbia Arts Council.

This book has been published with the help of a grant from the Canadian Federation for the Humanities and Social Sciences, through the Aid to Scholarly Publications Program, using funds provided by the Social Sciences and Humanities Research Council of Canada, and with the help of the K.D. Srivastava Fund.

Printed and bound in Canada by Friesens
Set in Adobe Garamond by Artegraphica Design Co. Ltd.
Copy editor: Dallas Harrison
Proofreader: Lana Okerlund
Indexer: Cameron Duder
Cartographer: Eric Leinberger

UBC Press
The University of British Columbia
2029 West Mall
Vancouver, BC V6T 1Z2
www.ubcpress.ca

*To George and Janet Stunden,*
WHO WERE THERE FOR THE BEGINNING OF THIS PROJECT.


*To Sean, Danica, and Alison Gouglas,*
WHO WERE THERE FOR ITS END.


*And to Peter Bower and Nancy Stunden,*
WHO HAVE BEEN THERE ALL ALONG.

# Contents

List of Illustrations / ix

Foreword / xi
*Graeme Wynn*

Acknowledgments / i

Introduction: The Wet Prairie / 1

1 Drains and Cultural Communities: The Early Years of Manitoba Drainage, 1870-1915 / 19

2 Jurisdictional Quagmires: Dominion Authority and Prairie Wetlands, 1870-1930 / 50

3 Drains and Geographical Communities: Experts, Highlanders, and Lowlanders Assess Drainage / 76

4 International Bioregions and Local Momentum: The International Joint Commission, Ducks Unlimited, and Continued Drainage / 110

5 Permanence, Maintenance, and Change: Watershed Management in Manitoba / 139

Conclusion: Chequer Board Squares in a Dynamic Landscape / 164

Appendices / 172

Notes /175

Selected Bibliography / 209

Index / 234

# Illustrations

FIGURES

1  Landforms of southern Manitoba / 5

2  Men and horses digging a drain / 27

3  Boating in a drainage ditch near Woodside, Manitoba / 30

4  Drainage districts in Manitoba, 1933 / 34

5  Posing with ditch-digging machinery / 35

6  The Mennonite West Reserve, Drainage District No. 12, and the Rural Municipalities of Morris, Rhineland, and Montcalm / 40

7  A dredge in a flooded area / 43

8  Map of Manitoba showing provincial government lands for sale, 1900 / 62

9  The northwestern interior of North America in 1894 / 65

10  Government survey party involved in drainage work, June 1911 / 79

11  Township plan, township XIII, range 1 east / 80

12  Map of Drainage District No. 12 / 81

13  Southern Manitoba showing elevation of drainage districts / 88

14  Municipalities and drainage districts, 1933 / 92

15  Double dyke drain / 102

16  Roseau and Souris Rivers / 114

17 The excavation of a drain in Drainage District No. 8 / 123

18 Big Grass Marsh in relation to the Mississippi flyway / 126

19 Ducks Unlimited publicity material / 129

20 Early conservation districts in Manitoba / 159

TABLES

1 Range of deviation from the average growing season precipitation / 7

2 National origins of residents of rural municipalities involved in the Red River Valley Drainage and Improvement Association according to the 1921 census / 41

A.1 Annual capital expenditures on drainage, approximate to nearest thousand, 1896-1932 / 172

A.2 The status of drainage districts in 1934 / 173

# Wetland Elegy?

*by*
*Graeme Wynn*

Aldo Leopold knew a thing or two about wetlands. He once thought to call the collection of essays that his publishers titled *A Sand County Almanac* a "Marshland Elegy," and the first of his "Sketches Here and There," which make up the second part of the book, entered the world wearing that label.[1] Here are its haunting opening lines:

> A dawn wind stirs on the great marsh. With almost imperceptible slowness it rolls a bank of fog across the wide morass. Like the white ghost of a glacier the mists advance, riding over phalanxes of tamarack, sliding across bog-meadows heavy with dew. A single silence hangs from horizon to horizon.

Silence ... and foreboding. Even as the quiet was broken by the dawn chorus, Leopold observed that "a sense of time ... [lay] thick and heavy" on this small part of Wisconsin.

> The peat layers that comprise the bog are laid down in the basin of an ancient lake ... These peats are the compressed remains of the mosses that clogged the pools, of the tamaracks that spread over the moss, of the cranes that bugled over the tamaracks since the retreat of the ice sheet.

An "endless caravan of generations" had contributed to the making of this place, so that the cranes whose clangor rang out over the marsh each spring stood, "as it were, upon the sodden pages of their own history."[2]

Pages of Leopold's "Sketches" later – after accounts of sand counties, wild rivers, farms, wolves, pigeons, cheat grass, and other things in places as diverse as Wisconsin, Illinois and Iowa, Arizona and New Mexico, Chihuahua and Sonora, and Oregon and Utah – Leopold came to reflect on another marsh, in Manitoba. This was Clandeboye. Leopold's companions on his visit there in 1941 found little at which to marvel. The marsh may have been "lonelier to look upon and stickier to navigate than other boggy places," but they failed to recognize its secret message. The mysterious western grebe that found refuge among Clandeboye's reeds was no more worthy of attention than any other bird: to their minds it was just another name to be checked off in the bird list, as its alluring "tinkling bell" call was jotted down (Leopold lamented) in "syllabic paraphrase" as "'*crick-crick,*' or some such inanity." For Leopold, by contrast, Clandeboye was "a marsh apart, not only in space, but in time" and a place that begged translation and understanding.[3]

Clandeboye Marsh, northwest of Winnipeg, on the southern edge of Lake Manitoba, is part of the Delta Marsh, at 15,000 ha the largest freshwater marsh in the Canadian prairies. Early in the twentieth century, this area accommodated a number of hunting lodges built by local businessmen and wealthy people from afar (including English royalty and Hollywood leading man Clark Gable). In 1941 when Leopold visited, water levels were relatively low, but the ten- to fifteen-year cycle of rising and falling waters in lakes and marshes that marked the early twentieth century (and many before) was essential to maintaining the vitality and diversity of marsh habitats.[4] Leopold knew this, and he was well aware that Clandeboye was, even then, a remnant in the larger scheme of things. "Uncritical consumers of hand-me-down history" might think that 1941 "arrived simultaneously in all marshes," he wrote, even as he reserved a different understanding for himself and the birds. Coming upon Clandeboye *they* knew that they had fetched up in "the geological past, a refuge from that most relentless of aggressors, the future." Yet unlike the grebe and the mink, and the wrens, ducks, herons, and falcons that animated the marshland, Leopold feared that this marsh would shrink like countless others under the relentless onslaught of "dredge and dyke, tile and torch" that had helped farmers drain and dry thousands upon thousands of hectares in the cause of "improvement." Some day, he reflected, even Clandeboye, "dyked and pumped will lie forgotten under the wheat, just as today and yesterday will lie forgotten under the years."[5]

Clandeboye lies squarely within the Manitoba landscape that is the focus of *Wet Prairie,* Shannon Stunden Bower's fine and innovative account of people, land, and water in agricultural Manitoba; and her study, like Leopold's brief commentary on Clandeboye Marsh, offers important insights into the course of prairie development. Where Leopold treats Clandeboye symbolically, as a transcendent encapsulation of large processes, Stunden Bower's work is both more deeply scholarly in its approach and broader in its spatial and temporal scope. Her work ranges beyond marshes and bogs, to consider the place of periodically, as well as permanently, wet land in the settlement of the west, and its arguments rest on prolonged and close-grained engagement with historical evidence. Strictly speaking, it is not an elegy (a mournful, plaintive poem), but an intriguing treatise on the consequences of the disjunction between settler aspirations and environmental circumstances in southern Manitoba. Still, Stunden Bower's quest to understand the development of prairie wetlands provides much food for thought to those who crave more than hand-me-down histories of Manitoba. Just as Leopold's essays heighten our appreciation of the world by bringing an ecological sensibility to observations of nature and a poetic sensibility to scientific description, Stunden Bower's chapters sharpen our vision of, and shift our perspective on, the development of lowland Manitoba, the wider prairie, and the Canadian state. She shares with Leopold a desire to get beneath the surface of things and to examine the world from several different angles, a commitment that Leopold scholar Daniel Berthold has compared with Nietzsche's notion of "'*Winkelübersehen*,' seeing around corners, or looking into the nooks of things."[6]

For Leopold, getting beneath the surface of things sometimes meant doing just that. Seeking to understand the grebes of Clandeboye, he buried himself "prone in the muck of a muskrat house," so that his eyes could absorb "the lore of the marsh," even as his "clothes absorbed local color." Dedicated student of the Canadian past though she is, Shannon Stunden Bower never went quite that far. Having come to her interest in the wet prairie through the disciplines of English, History, and Human Geography, however, she did deepen her knowledge of wetland science by immersing herself in an ecology field course based at the University of Manitoba's Delta Marsh Research Station, not far from the site of Leopold's earlier submersion. And just as the latter's efforts to observe young grebes "receiving instruction in the grebe philosophy" illuminated his Clandeboye essay, so Stunden Bower's commitment to developing a better understanding of the biogeophysical intricacies of marshes and wetlands allows her both to adopt unusual ways of looking at her native province, and to shed new light on the connections between environmental processes and social change by asking fresh questions of evidence in the archives and the landscape.[7]

The basic argument of *Wet Prairie* echoes a fundamental tenet of environmental history – that environmental and social forces interact in
innumerable, intricate ways to change both landscapes and human cultures.
This book offers a finely tuned narrative of back-and-forth interactions
– interactions understood to be both open-ended and indeterminate –
between human desires and environmental circumstances. More than this,
however, Stunden Bower's work offers a convincing demonstration that
environmental history is not simply, and merely, a distinct subfield of
scholarship fenced about by an insistence on nature's power to influence
the course of human history, but a wide-ranging, integrative approach to
the past capable of providing new perspectives on, and a more complete
understanding of, topics and themes long of interest to historians.

Central to all that follows in Stunden Bower's analysis is her recognition
that surface water is a particular form of mobile nature. This places her
analysis of the wet prairie in conversation with a growing body of work
on the communities of interest created by those elements of the natural
world that transgress human-made boundaries and challenge human conceptions of the proper order of things. As weeds spread, animals roam,
winds and soils blow, diseases diffuse, and waters flow, they cross fences,
defy attempts at control, create new niches, find their own levels, and
undermine ideas of ownership and responsibility based on assumptions
about the permanence, stability, and malleability of nature. Those people
affected by such challenges are often drawn to engage one another, as they
recognize the need to grapple collectively with problems that affect them
personally. But the terms of this engagement have to be worked out anew
in every case. Views of the issues involved – about causes and effects, about
rights and responsibilities, about strategies and solutions – differ. Efforts
to address them seem as likely to produce conflict as to lead to cooperation.
And thus problems of nature – or more specifically of the disjunction
between human expectations and natural processes – become crucibles of
community formation.

This is true at a range of scales. At the very local level, the old adage
reminds us, good fences (that keep my stock from your alfalfa) make good
neighbours. So too, as Stunden Bower demonstrates, "good" ditches did
not simply divert water from the wet fields of one farm onto an adjoining
property. Neighbours needed to talk and plan, and (ideally) cooperate; and
because drainage problems typically extended over considerable areas,
and attempts at solving them invariably involved transgressing ecological
and jurisdictional boundaries, many had to be engaged in the conversation.
In practice, such engagement was more easily facilitated through the

political process, in policy debates and voting booths, than by engaging district populations in a continuing conversation. Local and provincial governments, therefore, began to insert themselves between flooded fields and the dissatisfied farmers. But still there were costs to bear, and everyone's circumstances, needs, and aspirations differed. It was not easy to achieve consensus on the best course of action, and as so often in local politics during the nineteenth century, patronage helped to oil the wheels of action.

Scale up a notch and the problems are compounded. Much of southern Manitoba is flat, but water runs, naturally, downslope from the "naturally" higher terrain that flanked the wet prairie. Should those residents on higher ground be held responsible, in any way, for water that accumulated at the bottom of what Stunden Bower aptly characterizes as the physiographic "soup bowl" of southern Manitoba? Opinions differed, and for the most part they divided geographically. Again, resolution of this conundrum had to be worked out in the political realm as hydrologists and engineers argued the need to address flood problems at the watershed scale, settlers worried about the expansion of government implicit in the creation of new administrative districts, and politicians baulked at the costs, in dollars and votes, which might be incurred through any such action. Through all of this, as Shannon Stunden Bower points out, Manitobans with an interest in the drainage question "kept one eye on their government and the other on their land, evaluating each in the context of the other," to such an extent that the local environment affected the deployment of government authority and the implementation of ideological principles (page 168 below).

At yet another scale, wet prairie lands and the mobile water that rendered them so difficult to identify and map, shaped interactions between dominion and provincial governments in Canada and impinged on international relations between Canada and the United States. Through a series of complicated negotiations that are well reviewed in Chapter 2 of *Wet Prairie,* the federal government agreed to transfer wetlands (to which it retained title) to the Manitoba government in recompense for the latter's investment in draining and improving these tracts. But, with dominion or provincial ownership hinging on difficult questions about the wetness or otherwise of particular parcels of land (answers to which were affected by seasonal and longer term shifts in climatic conditions as well as by anthropogenic factors), coupled with scant understanding of the wet prairie environment in Ottawa's distant corridors of power, suspicion and confusion marked intergovernmental discussions, and ongoing environmental processes assumed considerable political significance.

Dealing with water flowing across property or township lines, or capriciously extending and shrinking the extent of "wet land," posed a significant set of challenges, but these challenges reached another order when mobile water ignored national boundaries. Early in the twentieth century, swamp drainage schemes in Minnesota ran excess water into the Roseau River, which flowed north across the forty-ninth parallel to join the Red River in Manitoba, and was – said distressed residents of the river's lower reaches – "too small in its bed to carry that mass of water" (page 116 below). Yet, where highlanders and lowlanders in Manitoba were often at loggerheads (largely because their upstream-downstream positions, and thus their interests, were fixed), conciliation shaped ensuing discussions about international flows of water (largely because rivers run north and south across the Canada-US border, and both nations prefer to find specific solutions to particular problems rather than invoke immutable upstream-downstream precedents). Here, as in much of the story of human engagements with the wet prairie recounted in these pages, local interests were balanced with those of the states involved, tempered by the advice of experts, and addressed with the relatively limited resources available. The cumulative consequence, as Stunden Bower notes, was to make the history of settlement and development in Manitoba's wet prairie "a collection of human-sized tragedies and triumphs, rather than an overarching narrative of power and exploitation" (page 10 below).

In detailing this collection of tragedies and triumphs through her singular focus on the challenges of wetland development, Stunden Bower opens a fascinating and important space for rumination on the role of the state in, and the ideological underpinnings of, Canadian society – past and present. This is no accident. Her work engages – indeed in some sense it might even be said to begin with – Ian McKay's argument, made a decade or so ago, that the history of Canada should be understood as a project of liberal rule.[8] In doing so, it picks up an interpretive thread, and joins a debate that has engaged a number of Canadian scholars in recent years.[9] For McKay, the rise and ultimate hegemony of liberalism in Canada (which he sees taking place between the mid-nineteenth and the mid-twentieth centuries), requires thinking about the nation's past "simultaneously as an extensive projection of liberal rule across a large territory and an intensive process of subjectification, whereby liberal assumptions are internalized and normalized within the dominion's subjects."[10]

At its most straightforward, McKay's claim is perhaps simply an affirmation that Canada is less a traditional (ethnic) nation than a product of liberal values or the triumph of "a certain politico-economic logic – to wit liberalism."[11] But read more profoundly it carries discussion into deep, dark, and even treacherous waters. Liberalism is a slippery word, and critics have lamented the imprecision of McKay's formulation of it as "*both* a set of *values* and a *project,* a universe of meaning and a will to act."[12] The "liberal order," some aver, is such a plastic concept that it contains "almost anything associated with what has traditionally been understood about modern developing societies, premised as they are on property relations of inequality and bourgeois individualism."[13] Moreover, there is little that is distinctly Canadian about the liberal order so conceived. In response, McKay has emphasized the value of a "workable and sensible definition that can enable useful conversations and shared insights," and it is in this vein that Stunden Bower engages McKay's work in the pages that follow, by exploring the implications of what she terms "colloquial liberalism" in Manitoba.[14]

Like generations of British North American settlers before them, and most of those who accompanied and followed them into the "old Northwest," many of those who came to Manitoba in the late nineteenth and early twentieth centuries harboured a strong desire to acquire a farm and the independence they believed land ownership would bring.[15] Widespread as it was, this desire was likely as much a reflection of experience, of long-remembered histories of insecurity, rising rents, evictions, and oppression among migrant families, as it was of theoretical or abstract principle derived from the writings of Thomas Hobbes, John Locke, and their ilk. Wherever its roots lay, however, the desire to own land spawned a strong sense of private property rights and a lasting affection for the virtues of hard work and self-reliance – of an attachment to some of the central values of liberal individualism – among successful newcomers.

Understanding this is of some help in understanding the conflicts produced by mobile water on the wet prairie. Talk of private rights tended to undermine necessarily more diffuse notions of shared responsibilities, and it quickly carried those engaged with mobile nature to an impasse: once water flowed off a particular parcel of land it was no longer the responsibility of the person who owned that tract; but when rising water spread across the lowlands, it trespassed on private property and encumbered its owner's right and capacity to use his land productively. Similar convictions shaped ideas about how drainage works should be organized and funded. The user-pay principle seemed most appropriate to those residents whose lands were dry: because successful drainage would increase the productivity and value of private property, it seemed self-evident that those who stood to benefit should bear the costs of improvement. But, those who lived in wet areas could, and did, argue that drains were not unlike roads and land surveys – necessary parts of the infrastructure required for successful settlement – that were widely and properly set in place by the state.

Here one confronts the complexities of nineteenth-century liberalism. For McKay, the liberal project is a pervasive and unitary order "in which liberty, equality, and private property are sanctified as the foundations of a civil society that valorizes individualism."[16] When the state appears in this formulation its mandate is to facilitate and protect the wealth-generating activities of individuals (by enabling them to establish title to property, by allowing them to get goods to market, by ensuring law and order so that they are not deprived of the fruits of their labours, and so on). So argued an earlier generation of economists and so, Stunden Bower suggests, the drainage works undertaken by the Manitoba provincial government were logical enough investments "for a liberal state concerned with facilitating capital accumulation" (page 11 below). Indeed they were. But might they have been more? Could other dreams find space alongside the market calculus? And, if so, what are the implications of seeing these types of investments simply and solely as strategies in the service of capitalism?

More than thirty years ago the historian of ideas, Allan Smith, thought it necessary to challenge the widely held view that Canada was a society built on deference, in which "belief in the rights of the community" trumped individualist doctrines.[17] It was time, he argued, to move beyond the notion that a "lack of enthusiasm for doctrines espousing the primacy of the individual" had led Canadians "to structure a transcontinental nation dedicated to the furtherance of essentially conservative aims."[18] Against American sociologist Seymour Martin Lipset's clever assertion that "Horatio Alger has never been a Canadian hero," Smith mustered a great deal of evidence that "the myth of the self-made man informed no small part" of English Canadians' thinking about their society in the sixty or seventy years before the First World War.[19] Although agricultural historian Vernon Fowke had, earlier, demonstrated that "pioneer agricultural self-sufficiency ... has been and remains a persistent Canadian myth," Smith argued, convincingly, that commentators' descriptions of the developing country were tinged with a strong and particular individualist hue.[20] Catherine Parr Traill voiced a common sentiment when she wrote: "In Canada, persevering energy and industry, with sobriety, will overcome all obstacles."[21]

Today the consensus is clear. Liberal individualism was a fundamental tenet of early Canadian development. But, as Quebec historian Jean-Marie

Fecteau has pointed out, "in liberalism the 'individual' is both a fundamental premise and an eminently problematic category, open to an
enormous diversity of interpretations."[22] In nineteenth- and early twentieth-
century practice, liberal values were defined and arranged in different
ways in different circumstances. As Stunden Bower notes, "the mismatch
between liberalism (with its emphasis on individuality and private property) and the wet prairie environment (with its water flows that continue
in disregard of property boundaries) prompted ... [efforts both] to redesign
the landscape and [to] reimagine liberalism." Flooding and drainage, she
avers, affected how Manitobans worked their land, how they related to
state, and "how they thought of themselves and their neighbours" (pages
168 and 12 below).

Within liberalism, there was also and always significant space for the
state to regulate and limit individual liberties and aspirations in the interests
of the common or collective good (defined in more than monetary terms);
and there were, perhaps increasingly frequent, instances of governments
using their authority and resources to protect the vulnerable. Then, too,
as Fecteau has shown in his recent work on crime and poverty in Quebec,
individualism is often tempered by mutualism and the development of a
shared sense of identity and responsibility.[23] This was the case even in the
United States where liberal individualism has been much celebrated over
the years.[24] So Stuart Blumin wrote in an impressive 1976 study of early
nineteenth-century Kingston, New York (where the population increased
from 3,000 to 16,000 in forty years), that this was "a town that moved
toward community as it grew toward urbanism."[25] So too in Stunden
Bower's account of Manitoba's wet prairie, the individualism associated
with prairie farming conflicted with the realities of regional development,
as the planning of drainage districts conceptualized the landscape in ways
that "prioritized environmental conditions over [private] property boundaries" and prompted lowland Manitobans to work together in addressing
the challenges they faced (page 82 below). Moreover, because flooding
occurred without regard for human geographies, responses to it helped
create communities of shared interest across settlements otherwise divided
by cultural, ethnic, or religious affiliations.

These, it seems to me in conclusion, are important contrapuntal notes
to set against liberal individualism's dominant key, with its emphasis on
private property rights. By drawing them to our attention, Stunden Bower
reminds us of the importance of paying careful mind to particular historical
circumstances and suggests to me, at least, the dangers inherent in, too

simply, equating liberalism with capitalist market principles. Defining people through their relations to property, as mere consumers of utilities or as "bundle[s] of appetites demanding satisfaction," insisted the great Canadian political theorist C.B. Macpherson, offering a critique of this tendency in classical liberalism, leaves "an impoverished view of life, making acquisition and consumption central and obscuring deeper human purposes and capacities."[26]

For Macpherson, the liberal tradition was much richer than this. Rightly understood it encompassed a humanistic democratic strain, which was typically overlooked by those – such as economist Milton Friedman – who regarded competitive capitalism as the foundation of economic and political freedom and for whom "freedom of the individual or perhaps of the family" provided the ultimate yardstick of the liberal condition. Macpherson was scathing in his assessment of *Capitalism and Freedom,* describing it as "an elegant tombstone of liberalism."[27] But it is Friedman's ideas that hold sway, permitting market forces to penetrate new, previously protected, realms of existence, allowing the regulatory and welfare functions of the state to be rolled back in the name of neoliberal governance, and confounding efforts to limit human impacts on nature by the rhetorical foregrounding of opportunity, growth, and individual choice. Distort though they do the ideas of a long line of political philosophers, and weak as the logic of some of their claims may be, these ideas resonate because they simplify the complexities of earlier relations between people, institutions and environments, because they have been naturalized to the point that they have become "common sense," and because they tap into a collective, albeit generally poorly understood, intellectual and social heritage. There can be no better argument for taking the past and the ways in which we represent it, as well as the careful arguments of this important book, with the utmost seriousness, as we grapple with the task of shaping a humanistic, civil society for the twenty-first century.

# Acknowledgments

I f I have learned anything over the years I have spent completing this book, it is that big projects take a lot of help. I am very fortunate to have had some of the best available guidance and support. *Wet Prairie* began in the Department of Geography at the University of British Columbia. I am very grateful to Matthew Evenden and Tina Loo, both of whom provided insightful criticism and valuable suggestions. In addition to careful commentary on my work, Gerald Friesen offered good advice and generous support, just as he has done since I was fortunate to find myself in his introductory Canadian history course at the University of Manitoba well over a decade ago. Graeme Wynn provided constructive criticism, careful guidance, and ample encouragement, from the half-baked idea stage through to the publication of this book. I am grateful to have had the opportunity to work with him. It was also a privilege to have David Breen, Cole Harris, and Nancy Langston engage with my work.

At the University of Alberta, I have benefited from the sage advice and careful criticism of my postdoctoral adviser, Gerhard Ens, as well as from the warm welcome and intellectual fellowship offered by other faculty members and students. I am especially grateful for assistance and opportunities extended to me by Sarah Carter and Paul Voisey. At both the University of Alberta and the University of British Columbia, I am grateful for the able assistance of support staff and departmental administrators. Many years ago I was an MA student in the Department of History at the University of Toronto. During my time there, I benefited from the support and guidance of Allan Greer, Sean Hawkins, and especially Paul

Rutherford. These scholars helped me to keep on the academic track and gave me opportunities to deepen my interest in the subfield of environmental history.

I would like to thank Geoff Cunfer for his support and understanding at a difficult moment in my academic career as well as for his generosity in sharing materials related to the study of the wetter portions of the North American Great Plains. I am grateful for the assistance of Dale Wrubleski (Ducks Unlimited) and Gordon Goldsborough (University of Manitoba Department of Biology and Manitoba historian). Scott St. George (formerly of the Geological Survey of Canada, now at the University of Minnesota) was very helpful. Over the years I have spent working on this book, I have learned a great deal from Emily Jane Davis, Arn Keeling, Josh MacFadyen, Melanie Niemi-Bohun, Jonathan Peyton, Liza Piper, Joanna Reid, and John Thistle. I have also appreciated Ashley Carse's willingness to share resources on matters of mutual interest. The feedback from the anonymous reviewers for UBC Press was invaluable, as were the support and guidance of the editors and production staff with whom I worked. I am grateful for the services and advice of cartographer Eric Leinberger.

I am thankful for the support (and the distractions!) provided by numerous good friends scattered across Canada and even around the world but concentrated in Winnipeg, Edmonton, and Vancouver. You know who you are.

For financial support during my time at the University of British Columbia, I am grateful to the university for the Tim and Ann O'Riordan Fellowship for Research in the Field of Sustainability, to Green College for a graduate fellowship, and to the Social Sciences and Humanities Research Council of Canada for its doctoral fellowship program. The process of writing this book stretched over my years at the University of Alberta, first as Grant Notley Postdoctoral Fellow and then as a Social Sciences and Humanities Research Council of Canada Postdoctoral Fellow. I am glad to acknowledge the financial support I have received in my positions at the University of Alberta. This book has been published with the help of a grant from the Canadian Federation for the Humanities and Social Sciences, through the Aid to Scholarly Publications Program, using funds provided by the Social Sciences and Humanities Research Council of Canada.

I am grateful for the assistance of archivists and librarians at Library and Archives Canada, the Manitoba Legislative Library, the Saskatchewan Archives Board, the University of Alberta Archives, and especially the Archives of Manitoba. The Network in Canadian History and Environment

(NiCHE) has done a great deal to support the subfield of environmental history in Canada, and I am grateful for the opportunities it has made available to me. I would like to thank *Environmental History* (published by Oxford University Press) for permission to reprint portions of "Watersheds: Conceptualizing Manitoba's Drained Landscape," *Environmental History* 12, 4 (2007): 796-819.

Family has been a constant source of support from the beginning of this project through to its end. Thanks to many members of my extended family as well as various members of the Gouglas clan. I am grateful to my sisters, Kirsty Howatt and Sarah Bower. My maternal grandparents, George and Janet Stunden, are remembered with love. Peter Bower and Nancy Stunden, my parents, have gone above and beyond in their support of this project, just as they always have done with my various undertakings. I am also grateful for their intellectual interest and good advice. Finally, I'd like to thank my partner Sean Gouglas, for so many reasons best known to us alone.

# Wet Prairie

# The Wet Prairie

In May 1889, prospective settler Alex Ingram wrote a letter to the Manitoba minister of public works. He had arrived in the province some months earlier "in search of land to settle upon." His intention was "to make a permanent home to my self." In this way, Ingram formed part of a wave of immigration following the Canadian government's assumption of control over a large swath of northwestern North America, the lands now known as the Canadian Prairies or northern Great Plains. The federal government solicited newcomers to help assert its claim to these lands in the face of an expansionist United States of America on the one hand and long-resident Aboriginal people on the other and with an eye to the creation of a political and economic empire across the region. Ingram and others who shared his ambitions arrived in search of available land in what was presented as a rich and only sparsely inhabited territory.

In an area near Roseisle, a small settlement west of the Red River and south of the Assiniboine River, Ingram found a soil that "was all that could be desired." For all the area's initial promise, however, a few wet spring weeks revealed a problem. What Ingram described as a creek without banks flooded the land and rendered it "unfit for cultivation." He was quick to recognize this was not a singular occurrence: it seemed clear to him that "the ½ section that I was thinking of taking will be useless in wet seasons." Despite the fact that "in many respects this is one of the most desirable localities in the Province," Ingram was convinced that if circumstances remained unchanged he would be obliged "to seek a home [some]where else."[1]

Alex Ingram had arrived in the wet prairie. His experience there was far from unique. This book is the story of the countless Manitobans like Ingram who found their agricultural ambitions in conflict with the region's environmental realities. The lands over which the Canadian government assumed control included a large, rather flat, often dry expanse that stretched like a tabletop from the Manitoba Escarpment west to the Rocky Mountains. But in an area extending north from the international border to the southern end of Lake Winnipeg and west to halfway along Lake Winnipegosis (an area described in various places in this book as southern Manitoba or agricultural Manitoba or, more precisely, Manitoba's wet prairie), the situation was different. There the land was more large soup bowl than tabletop, with slopes seemingly designed to collect precipitation and relatively impermeable soils that ensured the water pooled.[2] In many years, during the time when farmers most needed to be working ready fields, areas of the soup bowl were far too wet to farm. This book asks questions about agricultural life in Manitoba's wet prairie. How did residents perceive their surface water problems and those of their neighbours? In what ways did they attempt to alter what they saw as an unsatisfactory landscape? What happened when environmental change did not add up to the landscape they had envisioned?

This book is also the story of the Manitoba state's efforts to alter the region's environmental realities in favour of permanent and prosperous agricultural settlement. Early on, the province took charge of a significant effort to drain areas afflicted by excess surface water. Over the years, this effort evolved from digging a small number of key drains to creating a district system charged with managing the financing and engineering of intricate drainage systems. Successive political administrations were flooded by the complexities of surface water management: the problem of determining what was to be done and how best to do it; the competing claims of various experts and the dissenting opinions of local residents; the shifting parameters created by actions of other governments and evolving scientific thinking; and the challenge of coping with Manitobans' compounding fury when lands remained wet. These complexities had consequences: as the landscape was altered, so the government itself was affected. This book examines efforts to render the wet prairie landscape more suitable for agriculture and the patterns of conflict and cooperation that emerged early on and changed over time.

Alex Ingram's story has not been told before. Despite the prominence of agricultural themes in both Canadian and environmental history as well as the substantial amount of Manitoba farmland affected by excess water, the story of drainage in the province has attracted only a little scholarly attention.[3] To keep stories such as Ingram's at the centre of the narrative, this book differs from the many others that address prairie agricultural topics. It says little about the types of crops planted in the province or about other matters basic to any history of farming. In place of plowing techniques and commodity prices and transportation problems, this history addresses water flow patterns and engineering expertise and public opinion. What emerges is the story of the thinking behind and debates over land drainage, a particularly important, particularly complicated, and particularly contentious public project.

Most scholars concerned with the prairie environment have focused on the important issue of drought. Surface water was not hugely more abundant in Manitoba than in Saskatchewan and Alberta. For southern Manitoba, the mean annual total rainfall ranges from about 300 to 600 millimetres (12 to 24 inches). In most parts of southern Alberta and Saskatchewan, it ranges from about 200 to 300 millimetres (8 to 12 inches).[4] But even this relatively small measure was enough to make a significant difference, especially given the soup bowl character of much of agricultural Manitoba.[5] Overall, Manitoba was less prone to drought and could produce a greater variety of crops than the other prairie provinces. But it was more at risk of flooding. While scholars have suggested how the fortunes of the prairie provinces have diverged in recent decades, focusing on Manitoba's wet prairie reveals an earlier history of provincial difference rooted in the province's particular environment.[6] The wet prairie was a distinctive landscape, with distinctive challenges for those looking to farm it. This book, then, is the story of a landscape that Manitobans with their provincial government have long sought to alter and that scholars have not yet examined in detail. It is the history of a region that, while profoundly changed, remains persistently wet.

## PHYSICAL GEOGRAPHY

What is the wet prairie? The term "wet prairie" was used by nineteenth-century settlers to describe parts of the north-central United States that were, as later observers would determine, ecologically similar to much of southern Manitoba.[7] Although adopted early on by the United States Soil Survey, the term has fallen from use in recent years.[8] This is due in part to the development-focused American federal government's decision to exclude the wet prairie category from wetlands classification systems, thus reducing the amount of land subject to protective legislation.[9] A number of historical works addressing the wetlands of the United States in general, or focusing on the wet prairie region in particular, suggest the variety of political, social, cultural, and legal pressures bearing on the process of classifying some areas as wetland.[10] In this book, the term "wet prairie" refers to agricultural areas in which settlers often wrestled with what they believed to be too much surface water. The term provides a way of referring to the landscape without introducing distinctions that were meaningless to drainers, such as those between ecologically dissimilar wetlands such as bogs and swamps. It also incorporates the understanding that to identify any land as "wet" or "dry" is anthropocentric. Ultimately, the land simply is as it is – whether it is wet, dry, or just right is a human judgment.[11] Settler Alex Ingram was clear in his assessment of the land he wanted: it was too wet, and something needed to be done. The term "wet prairie" is used to refer to what he and his fellow settlers perceived as a problematic landscape.

Ingram's problematic landscape lies within the physical geography of agricultural Manitoba. In rough terms, the provincial south comprises three distinct landforms, oriented north-south: the Precambrian Shield, the Manitoba Lowlands, and the Southwest Uplands beyond the Manitoba Escarpment (see Figure 1). The lowlands have been described as the first prairie level and the uplands as the second prairie level.[12] The igneous rock of the Precambrian Shield, also known as the Canadian or Laurentian Shield, extends across northwestern Ontario and into Manitoba. This is, almost everywhere, a rough-hewn landscape, carved up by rivers, dotted with lakes, and covered with boreal forest. In southeastern Manitoba, the land generally slopes down to the west, with the transition to the lowlands occurring around the 300-metre (980-foot) contour line.

Once dominated by tall-grass prairie, the Manitoba Lowlands have proven to be an especially rich agricultural area. Varying from 64 to 80 kilometres (40 to 50 miles) in width, the lowlands are bisected by the northerly flowing Red River. The land slopes toward the river as well as toward Lake Winnipeg, though at very modest rates in both cases. Indeed, the flat, low expanse is known colloquially as the Red River valley, though the river had little to do with the creation of the lowlands. Rather, the valley is a remnant of glacial Lake Agassiz, an immense water body that dominated the continental interior for many years before warming temperatures prompted glaciers to recede, causing the lake to drain into the oceans. Gene Krenz and Jay Leitch describe the Red River valley, including the areas both north and south of the international border between Canada

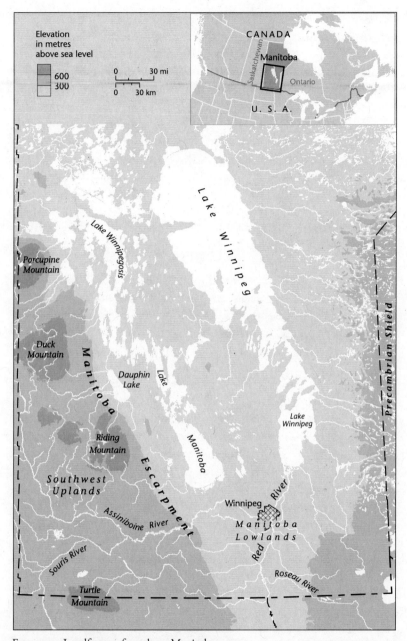

FIGURE 1   Landforms of southern Manitoba

*Source:* Adapted from Figure 2.9, "Physiographic Regions of Southern Manitoba," in M.
Timothy Corkery, "Geology and Landforms of Manitoba," in *The Geography of Manitoba:
Its Land and Its People,* ed. John Welsted, John Everitt, and Christoph Stadel (Winnipeg:
University of Manitoba Press, 1997), 21.

and the United States, as "one of the largest, truly flat landscapes in the world (roughly the size of Denmark)."[13] In many areas of the ancient lake bed, comparatively flat land and clay-based soils contribute to significant drainage problems. Above the western 260 metre (850 foot) contour line, soil permeability improves, though the land continues to rise at only a moderate rate.

The Manitoba Lowlands are bounded to the west by a southeast-northwest-trending escarpment, which is marked by a rapid increase in elevation. Beyond the escarpment are the Southwest Uplands, where the land is more undulating and the precipitation more limited. Over the past century, fire suppression and the absence of grazing bison have allowed aspen and bur oak to invade the uplands, with rapidly growing scrub cover moving south from the boreal forest that extends across the northern Prairies.[14] This has resulted in an expansion of the parkland, a transitional area between forest and grassland, even as more and more land has been cleared for agriculture. The Assiniboine River flows east through the Assiniboine Delta, a preglacial embayment that interrupts the scarp face to the south of Lake Manitoba. Diminishing rates of precipitation to the west are reflected in the river's flow. Although the Assiniboine drains about 101,388 square kilometres (39,146 square miles), its average flow is much smaller than that of the Red River (which has a drainage basin of only 77,249 square kilometres [29,826 square miles], when the Assiniboine's contribution is set aside).

Taken together, these three distinct areas form the macro-topography that affects surface water patterns in Manitoba's wet prairie. They are the sides and bottom of the Manitoba soup bowl. Awareness of the changes in elevation that distinguish them and of the key geographical features of each takes us some distance toward an understanding of what would be perceived, beginning in the later years of the nineteenth century, as Manitoba's drainage problems. The following chapters examine specific instances of flooding attributable not only to this basic physical geography but also to the local topography nested within it and to the climatic disturbances that moved over it.

The wet prairie was an extremely dynamic landscape, due in large part to variability in both precipitation and temperature regimes.[15] Variability operated geographically, creating a checkered landscape of wet and dry, as well as temporally, with some periods far wetter than others. The landscape's dynamic character is suggested by Table 1, which provides the range of deviation from the long-term average growing-season precipitation at three

TABLE 1

**Range of deviation from the average growing-season precipitation**

| Years of record | Place | Average growing-season precipitation | Standard deviation | Percent of average |
|---|---|---|---|---|
| 62 | Winnipeg | 244.35 mm (9.62") | 78.74 mm (3.10") | 32 |
| 32 | Cypress River | 208.03 mm (8.19") | 72.90 mm (2.87") | 35 |
| 51 | Morden | 216.66 mm (8.53") | 90.17 mm (3.55") | 41 |

*Note:* This table lists rates of growing season precipitation as opposed to total annual precipitation as specified for southern Manitoba earlier in this chapter.
*Source:* Modified from A.J. Connor, *The Climate of Manitoba* (Winnipeg: Province of Manitoba, Economic Survey Board, 1939), 12.

places in southern Manitoba: at the City of Winnipeg (around the forks of the Red and Assiniboine Rivers), at Cypress River in the Rural Municipality of Victoria (about 155 kilometres west and slightly south of Winnipeg), and at Morden in the Rural Municipality of Stanley (about 130 kilometres southwest of Winnipeg). For all three locations, the growing-season precipitation varies from year to year by an average of over 30 percent. In creating the distinctive wet prairie environment, variability in rate of precipitation was more important than simple quantity.

In this dynamic hydroclimate, the risk of flooding at any given time or place was defined in part by soil moisture content, winter snowfall, melt patterns, and spring rain. Vulnerable Manitobans kept a close watch on the interaction among these key variables, not only in their immediate surroundings but also over a significant portion of the Saskatchewan-Nelson River basin. This immense drainage system discharges through Hudson Bay, and much of the water has first flowed through the provincial south. There is some truth to the assertion that Manitobans sit "downstream from everyone else," and increased flow from the east, west, or south has contributed to flooding.[16] Change was basic to the environment of southern Manitoba, and water was a primary driver of change. Variability in water patterns should not, however, be regarded simply as a threat to regional well-being. Wetland ecologists have made it clear that expansion and contraction of wet areas are fundamental for continued ecosystem vitality.[17] Some of the province's most fertile soils are those in the Red River valley that have been enriched by deposition during periods of flooding.[18] But even as it helped to create a rich soil in which it was easy to envision agricultural success, environmental variability was not conducive to what newcomers understood as agricultural progress. For example, settler Alex Ingram, who complained about good land ruined by inundation, did not connect the richness of the soil to its susceptibility to flooding.

Given the post-1950 frequency of inundations large enough to threaten Winnipeg, the province's capital and largest city, it is hardly surprising that mention of excess water in southern Manitoba tends to evoke thoughts of catastrophic flooding along the Red or Assiniboine River. Expanses of water spreading back across farm fields, Winnipeggers labouring on sandbag dykes, and evacuated flood victims nervously watching the weather: all are among the well-known contemporary images of the region.[19] Catastrophic flooding has been a serious concern in Manitoba, and it is considered briefly in some of the book's later chapters. But by and large, the analytic focus remains on the problem of agricultural flooding, which was far more significant to the development of the province.

## Scholarly Landmarks

The history of Manitoba's wet prairie connects with a number of scholarly landmarks. Given the provincial government's important role in land drainage, scholarship on the state is particularly significant. State formation has been a key concern of Canadian historians in recent years, with legal, managerial, and economic processes seen as fundamental to the establishment and extension of government authority. Many historians interested in the state have been particularly concerned with power and how different social groups (such as those defined in terms of race or gender) had different rights and responsibilities. Others have been concerned with mechanisms of regulation, examining law enforcement, the census process, and the school system.[20] Despite the varied emphases of these scholars and the overall sophistication of their work, few have accorded much attention to the diverse landscapes over which the process of state formation played out. Only the bare fact of distance has received much attention, with technologies of administration such as the census understood as useful in part for how they facilitate governance from afar.[21] Historian Tina Loo is one exception, as her work on the legal system in British Columbia examined how mountainous topography complicated justice delivery.[22] Loo's more recent study of conservation in Canada, in a manner typical of much work in the growing field of Canadian environmental history, emphasizes actions by the state over effects on the state.[23] Particularly given the environmental differences within the vast Canadian landscape, there is room for much further scholarship that explores how the development of the state related not only to the people but also to the landscapes that it incorporated.

In Manitoba, surface water was a local environmental factor that inflected how provincial authority was extended and entrenched. It affected when, where, and how state authority was deployed. Because drainage was expensive and had to be coordinated at a level beyond the individual farm, early on the Manitoba state interjected itself between the flooded field and the dissatisfied farmer. Considering the emergence of a system of special district government dedicated to surface water management as well as generations of lively debate over the province's role in all of this, surface water management bore on the ultimate shape of the government itself. Environmental historian Donald Pisani has called "the proliferation of special districts" such as drainage districts "the untold story" of natural resources administration in the twentieth century.[24] The history of drainage in Manitoba, then, speaks not only to those interested in scholarship on the state in a Canadian context but also to an international academic community concerned with the administration of natural resources.

An important thread in this international scholarship focuses on relations between the state and irrigation. Although it would be an oversimplification to identify drainage and irrigation as the inverse of the other, the two undertakings do share challenges related to water management, infrastructure construction, and public administration. In 1957, Karl Wittfogel proposed that the construction and operation of large-scale irrigation works contributed to the emergence of despotism among what he termed hydraulic societies.[25] Since that time, numerous scholars have built on or reacted to his thesis.[26] For environmental historians of North America, Donald Worster's analysis of irrigation in the American west represents a compelling illustration of the state in service to capital. The hydraulic society Worster saw was defined by "a coercive, monolithic, and hierarchical system, ruled by a power elite based on the ownership of capital and expertise."[27]

Although there are basic similarities between state-assisted drainage in Manitoba and state-assisted irrigation in the American west, there are also important differences. A key difference is the division of power between experts or managers on the one hand and local residents on the other. By the early twentieth century in the American west, as Worster explains, rich investors or large agriculturalists were enlisting engineering experts to manage water for their own benefit. In Manitoba, a landscape where the prize was effective drainage on an individual's land rather than a portion of water transported from a perhaps-distant region, engineering expertise assumed a different quality. Manitoba farmers knew their own lands and were willing to challenge engineering experts and government administrators on the basis of their knowledge. Furthermore, through letter writing and collaborating with others and presenting at hearings, they first created and then were given opportunities to do so. This study provides a counterpoint to Worster's vision by illustrating how water management might actually prompt meaningful interplay among experts, the state, and its citizens. In Manitoba, this interplay was intensely local, involving a broad swath of the interested public and invoking the local landscape. If one basic difference between Manitoba and Worster's American west was the quantity of water, another was the quality of engagement between people and their government.

All of this might have been possible because of another key difference between Manitoba's wet prairie and the arid American west: the amount of money in circulation. There were no fortunes to be made in draining the farms of Manitoba, as was discovered by the few private companies that, for a brief time, attempted to profit from drainage in the province by acquiring wet land, undertaking drainage, and then selling dry land at inflated sums. Both the enduring uncertainty surrounding the reliability of drainage and the continued availability on good terms of land from the federal government likely contributed to the early failure of these endeavours. Much drainage in Manitoba was undertaken by local contractors with the modest tools needed to accomplish the small contracts let by the provincial government. Although livelihoods were certainly at stake, no one was becoming shockingly rich. This might be an important part of the reason the sort of dramatic language Worster uses in relation to the arid American west, invoking concepts of empire and domination, seems to be inappropriate in relation to the newcomer experience in Manitoba's wet prairie, however much the province's agricultural transformation might have been part of a process of colonialism. Here there is a collection of human-sized tragedies and triumphs rather than an overarching narrative of power and exploitation. At least in the early decades of the twentieth century, when much drainage in Manitoba took place, this was capitalism at a different scale.

A second scholarly theme important in *Wet Prairie* is the issue of liberalism. From Fernande Roy's 1988 analysis of the liberal orientation of Montreal's business class at the beginning of the twentieth century, through Barry Ferguson's 1993 intellectual history of influential political economists at Queen's University, to Tina Loo's 1994 study of the legal apparatus in mid-nineteenth-century British Columbia, liberalism has been in recent decades an important analytical concept for Canadian historians.[28] In its various forms, liberalism has been identified as the ideology underpinning the orientation of influential Canadian groups and institutions toward individualism, property ownership, self-determination, and capitalism. In 2000, Ian MacKay argued that, despite this significant scholarly attention, Canadian historians had not yet grasped the full significance of liberalism. In his view, the concept provided the basis for a reorientation of Canadian history toward the study of the nation as a liberal project. At the least, this was an argument for the continued importance of liberalism as an analytic concept as well as an invitation for debate with those who agreed on the importance of liberalism but differed on its particular presentation.[29] Influenced by works by Canadian historians as well as an older international scholarly tradition that fleshed out the intellectual and cultural origins of the liberal concept, recent monographs – including those by Ruth Sandwell (on the aliberal residents of Saltspring Island), James Murton (on the post-World War I soldier settlement programs in British Columbia), Catherine Wilson (on tenancy in nineteenth-century Upper Canada), Jarett Rudy (on smoking in Montreal), and Daniel Samson (on industry and improvement in Nova Scotia, 1790-1862) – have continued to ask important questions about the effects of liberal ideas on Canadian life.[30] It is Murton's work that is most relevant to an analysis of the Manitoba situation, as his study of environmental change in the BC countryside illustrates the value of studying in tandem Canadian liberalism and environmental history.[31]

*Wet Prairie* contributes to this scholarly tradition by examining what happened to liberalism when deployed in a wet and variable landscape. Under liberalism, it was the state's role to facilitate and protect the wealth-generating activities of its citizens. Transportation, law and order, and land surveying were all activities undertaken by liberal states to create an arena in which capitalism might operate more successfully. In the wet prairie, surface water was a serious and persistent threat to agricultural prosperity. The Manitoba government's drainage activities were a logical undertaking for a liberal state concerned with facilitating capital accumulation. This assessment is in line with that of economist W.T. Easterbrook, who situated drainage among the diverse array of strategies (ranging from road construction to farm credit programs) through which Canadian governments extended assistance to agriculturalists.[32] Through digging individual drains and then through coordinating the engineering and financing of substantial drainage systems, the Manitoba government sought to ensure the productivity of privately owned farms. Drainage was part of the infrastructure of settlement, similar in some respects to other under-takings by the provincial government as well as to undertakings by liberal governments located elsewhere.

Although drainage in some ways resembled other government under-takings such as road building, it was distinctive in a few important aspects. Drainage directly affected a large number of rural Manitobans, with local interest spread at least as wide as surface water flooding in the spring. Notably, this was not an interest that peaked at drain construction and then declined. As surface water management often remained a contentious issue even in the drained landscape, few became inured to which drainage had been undertaken, and many stayed engaged in ongoing debates over what should be done next. Both flooding and drainage served to unite Manitobans around shared experiences related to surface water manage-ment. Given the importance of state-managed drainage, communities of interest emerged largely to advocate more effectively for undertakings likely to bring significant widespread benefit or, later, to oppose financial involvement in schemes likely to bring little local gain. Such communities of interest overlay key divisions such as municipal boundaries, which in Manitoba often coincided with notable differences in ethnic concentration. In this way, these communities of interest reflected a distinctive way of conceptualizing the land and the people who lived there, a way that was based on vulnerability to surface water flooding. Flooding and drainage changed how affected Manitobans farmed, how they related to the govern-ment, and how they thought of themselves and their neighbours.

If liberalism is part of the explanation for why the Manitoba govern-ment undertook drainage, it is also part of the reason drainage was so controversial. Private property is a key liberal tenet. For farmers, independ-ent ownership of a parcel of land, even if this was only an aspiration, offered the opportunity for productive labour as well as an assurance of individual freedom. In the late nineteenth century and early twentieth century, many settlers came to Manitoba in the hope of achieving independent ownership of a farm. Likely, they did not give much thought to how water flow pat-terns would ensure their fortunes were intertwined with those of their neighbours. The steep sides and flat bottom of the Manitoba soup bowl meant that surface water ran from the lands of some (those on the sides) and pooled on the lands of others (those on the bottom). This soup bowl, then, collected not only water but also opinions about how land should be managed and drainage should be funded. Those on the sides thought they should in no way be responsible for the costs of dealing with the water that flowed off their lands: although various justifications were offered, their fervour stemmed from the conviction that this was a matter of private property. Why should they pay for drainage works serving lands they did not own? Those on the bottom thought it was utterly unfair for them to be saddled with the task of disposing of water that had flowed in from other lands: this was also a matter of private property, as water from elsewhere was encumbering their ability to make productive use of their lands. Why should they suffer under the excess water that ran down from lands over which they had no control? Both perspectives were examples of colloquial liberalism in Manitoba, of the ways abstract liberal principles were understood and deployed by non-experts in the context of a particular social, political, and environmental landscape.[33] They were the key poles in a long-running, bitterly fought battle over how the wet prairie should be drained and who should pay for drainage.

The third scholarly landmark significant to this book is actually a set of themes related to the challenge of managing a mobile resource such as water in a human world of borders, jurisdictions, and property lines. Environmental historian Mark Fiege has been concerned with the management of those elements of the natural world that cannot be easily made to respect property lines.[34] Unwanted plants, wild animals, and crop diseases are some examples. Plant seeds blow, animals roam, and agricultural plagues spread, all without concern for human attempts to divide up the farm landscape. Such movement creates what has been called an ecological commons, in which the private property landscape is overlaid by elements of the natural world that are of common concern to all landowners. In southern Manitoba, surface water was a particularly significant variety of mobile nature, creating an ecological commons among those touched by its flow patterns. Even liberal farmers preoccupied with their private property were interested in how others managed surface water since their own land might be affected by changed flow patterns. In the American west, a region divided into fiercely defended private holdings, Fiege illustrated that both conflict and cooperation emerged from efforts to adapt to environmental factors that extended across property lines. Similarly, in Manitoba's wet prairie, farmers were drawn into engagements with each other as they sought to address the surface water situation. This engagement persisted in the drained landscape, for even when drainage operated relatively successfully farmers remained profoundly interested in the flows of surface water. Ultimately, it was this engagement as much as the water itself that characterized Manitoba's ecological commons.

Just as flowing water bore on relations among landowners, so too it complicated dealings among governments. Difficulty often ensued from the way this mobile resource confounded efforts to establish permanent lines on the land and to govern according to these boundaries. This was evident between the province of Manitoba and the government of Canada as uncertainty over land condition, whether flooded or dry, affected negotiations over which government had the rightful claim to the lands in question. This was also the case between Canada and the United States as surface water flows across the international border amounted to an environmental management problem addressed through protracted negotiations. In addressing these issues as well as conflicts between adjacent rural municipalities and between rural municipalities and the provincial government, *Wet Prairie* provides an opportunity to consider at multiple scales, from the rural municipality to the province to the nation-state, the challenges to surface water management. It illustrates how a single environment, the wet prairie landscape, confounded people and governments in a variety of ways over a long period of time. *Wet Prairie* also makes clear how transboundary environmental management offered certain opportunities along with an array of difficulties. From progressive environmental management along the Canada-US border to the significant work of Ducks Unlimited, an international agency, new ideas and approaches often developed from engagement with interests outside Manitoba.

The ways *Wet Prairie* addresses the ecological commons, jurisdictional disputes, and transboundary environmental management issues provide the basis for an engagement with the concept of bioregionalism. If an ecological commons is the result of an environmental characteristic that creates a shared interest among human land users, a bioregion is a geographical entity created by those who manage or study the land in reflection of an ecological element particularly significant to the area's past, present, or future. Some environmental historians argue for bioregional history, an approach that defines the study area in terms that make sense environmentally rather than culturally or politically. For instance, Dan Flores asserts that environmental historians should derive their subjects of inquiry from ecological categories. In a much-quoted phrase, he suggests that "the politically-derived boundaries of county, state and national borders are mostly useless in understanding nature."[35] Environmental historians, Flores puts forth, should investigate the "natural nations" of the world – those defined by boundaries that "make real sense ecologically and topographically."[36]

In the nineteenth and twentieth centuries, historians, geographers, scientists, and popular writers concerned with southern Manitoba produced a number of works addressing an environmental unit that spans the international border, focusing on concrete geographic linkages such as a common glacial history, similar agricultural conditions, and the shared risk of catastrophic flooding.[37] These shared elements seem to be more than sufficient to fulfill Flores's criteria for a region that makes sense in ecological and topographical terms. So why not study the wet prairie as a natural nation that spans the forty-ninth parallel? This might seem to be sensible if the wet prairie alone were the subject of study; if the focus is also on the relations between the various governments and agencies responsible for its administration, then the matter is less clear. Although bringing the American states of Minnesota and North Dakota into play would present opportunities for comparative analysis, doing so would diminish the attention focused on the particular relations that developed in response to the Manitoba portion of the wet prairie. *Wet Prairie* is both more than and less than a bioregional history of the wet prairie, offering a great deal about human-environment relations in one portion of the bioregion but comparatively little about what went on south of the forty-ninth parallel. This study aims to contribute to the field of environmental history by illustrating the continued historiographical significance, even for environmental historians, of political boundaries.

## Change in the Wet Prairie

The chapters of this book trace the evolution of surface water management, emphasizing the particular perspectives that emerged in relation to provincial geography and illustrating changing approaches to life in the wet prairie. The first chapter lays out the problem of excess surface water as perceived by newcomers. It describes how the provincial government expanded to meet the demand for drainage and how the ecological commons prompted cooperation among culturally divided groups. Manitoba has a number of identifiable ethnic communities, including, for example, French, British, Ukrainian, and Mennonite. Many members of these communities live relatively close together in areas affected by surface water problems.

Chapter 1 briefly examines the Mennonite experience, using this group as an example of how affected Manitobans were prompted to work together on drainage. The chapter also addresses how Aboriginal people, who had a distinct relationship with the federal government, were poorly served when it came to drainage. Flooding on Aboriginal reserves did not threaten the larger process of agricultural expansion across the prairie. Although drainage districts were created in non-reserve areas to address the physical and administrative challenges of surface water management, neither federal nor provincial officials were prepared to make any significant effort to tackle the particular barriers to drainage of Aboriginal lands. The ecological commons was fractured by what was perceived as cultural difference.

The second chapter engages what is surely one of the oldest themes in Canadian history: prairie control of regional resources. Much political history of the Prairies is dominated by discussion of inter-regional colonialism, stemming in large part from the conditions under which Manitoba, Saskatchewan, and Alberta entered Confederation. Contrary to the precedent set in 1867 when the provinces of Canada (consisting of earlier versions of the provinces of Ontario and Quebec), New Brunswick, and Nova Scotia were united by the British North America Act, all three prairie provinces were denied control of the lands and resources within their borders. *Wet Prairie* offers something new to the study of an intergovernmental dynamic that remains basic to understanding many aspects of the Canadian past. My argument is simple: dominion-provincial relations, at least insofar as natural resources were at issue, were affected by environmental conditions in the wet prairie. Especially in the second chapter, I illustrate how an inadequate understanding of the Manitoba environment, in combination with the dynamic character of the wet prairie landscape, compounded the problems inherent to renegotiating the province's terms in Confederation. The history of the wet prairie suggests that, with respect to this one matter at least, the question of what should be considered within Canadian political history is no clearer than the borders of a typically variable and indistinct prairie wetland. In this way, *Wet Prairie* illustrates that environmental history is not simply an alternative view of the past but is rather a basic approach necessary for a more complete understanding of topics and themes already of concern to Canadian historians.

The history of state development and environmental change in Manitoba's wet prairie would be incoherent, of course, without reference to larger processes, and I consider some of them in the third chapter. As the patronage practices that dominated provincial politics through the 1910s fell into disfavour, the scientific expertise becoming increasingly prominent nationally and internationally provided an alternative logic for government decision making. In the case of drainage, the participation of scientific experts reflected a growing appreciation for the complexity of surface water management, even while buttressing overly optimistic hopes for successful resolution of flooding. At the same time, expert recommendations aligned with how some Manitobans perceived the drainage problem but conflicted with the views of others and thus served to complicate further the human landscape of drainage. A debate over values emerged as Manitobans tried out various ways of thinking about water management in the province. Interested parties deployed a variety of colloquial liberalisms, with the justification for an individual's assertion rooted in the geography of the individual's farm. Ultimately, one way that Manitobans evaluated the legitimacy of ideas about the role of the state was in relation to local environmental conditions.

Managing the wet prairie was challenging in part because it reached south into the United States, with water flowing across the international border. The first part of Chapter 4 considers how transboundary flooding along the Roseau River led Canadian and American officials in the late 1920s and early 1930s to endorse surface water management policies that seem, by contemporary standards, far more reasonable than those maintained by the Manitoba government at the time. The second part of the chapter addresses the human and environmental consequences of attempts to protect continental waterfowl populations. American waterfowl activists founded Ducks Unlimited, a conservation organization that remains in operation internationally, as a means of working toward the restoration of waterfowl habitat in Canada. But in neither the Roseau River nor the Ducks Unlimited case did innovative management of what were perceived as discrete environmental issues prompt wider change in surface water management across Manitoba. This chapter also illustrates how the drainage project had become entrenched in Manitoba, remaining the dominant approach to the wet prairie even as thinking evolved in relation to other environmental issues.

Eventually, in light of broad changes in thinking about economic matters, the public context for drainage discussions shifted. The fifth chapter examines these changes, beginning with a survey of ideas about the permanence of the drainage infrastructure. Extensive public funding became more acceptable in the post-World War II period. This helped to render viable administrative options that had been dismissed by many, with the provincial government dedicating substantial funds to environmental problems of personal concern to only some people in some areas of the province. At about the same time, new recognition of the problem of soil erosion by surface water expanded the community that perceived itself as directly interested in surface water management. Those occupying higher lands who had resisted any suggestion they were implicated in the flooding that afflicted lower areas became interested in watershed-based surface water management once the upper reaches of the watershed began to suffer the effects of rapid and increased runoff such as soil loss and slope instability. But shifts in agricultural economics and technology meant that little real change ensued. Highlanders were now willing to accept watershed management in theory but were concerned about putting it into practice at a time of profound change in the human landscape of the wet prairie. Despite opportunities related to new federal government programs leading to some efforts to resolve persistent environmental problems, the situation remained largely unchanged across many areas of the wet prairie.

A major barrier to progress in surface water management was the prevailing dynamic between government and residents, which had become profoundly negative in light of the failure to achieve the desired drained landscape. Dissatisfaction and distrust prevailed and severely circumscribed the possibility of cooperation between Manitobans and their governments. In *Seeing like a State,* James C. Scott set out to examine why some state projects failed.[38] *Wet Prairie* builds on Scott's analysis by examining the consequences of what was perceived as government failure in the Manitoba context. Falling short of the ideal of a permanently drained agricultural landscape had significant and long-lasting consequences for those involved in the administration of drainage. The profound irony beneath all of this was that, despite the accumulating complaints of wet prairie residents and the corresponding frustration of provincial administrators, continued flooding was largely a consequence of an extremely dynamic environment. It was an environmental reality, not the product of human negligence.

The history of drainage in Manitoba illustrates some of the consequences of rapid settlement by newcomers with a limited understanding of their new region. But this history also exposes how acquiring knowledge of a place was not a straightforward undertaking, not simply a question of the passage of time. Rather, it was a matter of struggle and debate, with concrete outcomes that affected the environment of the province. The process also had important consequences for both the people of the region and the governments that administered it. And this process remains ongoing, as those in the wet prairie continue the work of reconciling human ideas and environmental realities.

# 1

# Drains and Cultural Communities: The Early Years of Manitoba Drainage, 1870-1915

In the late 1850s, Henry Youle Hind explored the area that became the province of Manitoba as part of an assessment of the northwest on behalf of the Canadian government.[1] Heading westward from the southwest shore of Lake Manitoba, he traversed some of the most pronounced ridge and swale topography of the province. In his journal, he described how tiresome it was to "wade through marshes and bogs, separated by low ridges." In fact, Hind wrote, the land "may be said to be made up of marsh, bog, ridge, marsh, bog, ridge in most wearisome succession." This did not dampen his enthusiasm for the area. Drainage, he asserted, could transform the region: "I know of no other enterprise of the kind which could be executed with so little cost and labour and promise at the same time such wide spread beneficial results."[2]

Later provincial officials would certainly have laughed ruefully at Hind's projection, had it come to their attention. After but a few years of work, it was abundantly clear that the drainage necessary in many regions of Manitoba was both costly and labour intensive. Importantly, these difficulties were not entirely attributable to the wet prairie landscape traversed by Hind. The settlement system used on the prairie contributed to surface water problems. The federal government divided the land into square parcels without accounting for the surface water patterns of the wet prairie. These parcels were available to settlers at attractive terms, but many newcomers found their land was vulnerable to surface water flooding. Whereas the federal government was in charge of land settlement, the provincial government was responsible for public works such as drainage. This juris-dictional separation meant that surface water problems did not operate as a brake on the settlement process and left the Manitoba government to do what it could to assist individuals and communities in need of drainage. Many settlers were struggling with flooding, and the province as a whole was not thriving. The particular pressures bearing on Manitoba at this time rendered it all the more important that the province act to protect the wealth-generating capacity of its citizens. In this context, residents were intensely interested in government actions bearing on surface water, and what the province did or did not do often became fodder for intense and persistent debate.

And what about Hind's anticipated "wide spread beneficial results"? The second part of this chapter focuses on the human consequences of inundated lands on the one hand and drainage efforts on the other. The ecological commons of surface water extended across municipal bound-aries, property lines, and cultural differences. When faced with the prospect of being flooded off their lands, Manitobans from various cultural com-munities sought solutions to flooding. Building effective drains required cooperation with other interested parties and the provincial government. Beyond a changed environment, the result was the recognition of shared interests and the creation of parallel orientations toward the provincial state. The Mennonite experience provides an example of this for a particular cultural group. For Aboriginal people, the situation was very different. Although water certainly flowed across the lines between reserve and non-reserve land, Aboriginal people were cut off by barriers of jurisdiction and race from the drainage options available to other Manitobans. The third part of this chapter addresses how the human factors that figured in the drainage problem were different on Aboriginal reserves. Because of this difference, drainage served to further entrench the racialized distinction between Aboriginal and non-Aboriginal lands.

SETTLEMENT POLICY AND THE LOCAL ENVIRONMENT

Although wetlands exasperated travellers such as Hind, the same areas were valued by those with a lengthy history in the region. Native people had long used the northeastern prairie prior to European settlement, with different populations moving in and out of the area over time. Indigenous groups shifted locations of residence and provisioning partly according to environmental factors. Surface water conditions were among the many variables that affected community movement. Early European settlers joined Aboriginal and Métis people in wetland exploitation, gathering valuable plants and preying on abundant game.[3] Wetlands were important for a variety of reasons. They supported various useful plants and provided a needed source of water on the open prairie. They also attracted migratory waterfowl, and the ducks and geese in turn drew hunters. In these ways, the wet prairie proved to be an invaluable resource.

After the creation of the province of Manitoba in 1870, and especially with the subsequent influx of newcomers, land use practices began to shift. Residents certainly continued to utilize wet areas, but the newly established government authorities focused on the expansion of agricultural settlement. Early residents had not only taken advantage of opportunities presented by wetlands but had also accommodated the irregular geography of the wet prairie, building houses at dry elevations and pasturing animals in the slightly wetter areas that grew excellent hay. In contrast, agricultural settlement was to be defined not by local environmental conditions but by the grid-based land management system established in 1872 by the Dominion Lands Act.

This legislation, largely modelled on the American Homestead Act of 1862, specified the terms governing the alienation of the crown lands of the North American northwest.[4] Although the specifics of the arrangement were amended on a number of occasions, the general parameters endured through the settlement period. Those who met certain qualifications of age and gender were entitled to make entry on a homestead: that is, to establish a preliminary claim to a parcel of land. There was a small filing fee of ten dollars to pay immediately and conditions of residency and cultivation to satisfy over a period of years, but these were nevertheless terms designed to entice newcomers. The Canadian politicians who adopted the system hoped to spark a land rush comparable to that under way in the United States. Ultimately, settlement in both the Canadian and the American wests was part of what historian John Weaver has described as the international land rush that helped to define the modern world.[5]

To manage the land rush, administrators needed some system to carve up the northwest into homesteads, to enable the identification of a plot of land, and to record when homesteaders made entry. The township survey, a massive project of the Canadian government already under way by 1872, made it possible to pinpoint particular parcels of land in a vast landscape. It divided the northwest into large squares known as townships. Townships were numbered northward, with township one abutting the international border. In Manitoba, range referred to the distance east or west from the principal meridian, which ran north-south slightly west of what became the city of Winnipeg. For example, township one, range one west refers to an area of six miles square situated along the international border immediately west of the principal meridian. Each township was divided into sections of 640 acres, which were then divided into quarter-sections of 160 acres. These quarter-sections were available to newcomers.[6] The system of land alienation achieved the goals of facilitating administration and attracting newcomers, though it would take decades before the desired immigration rates were reached.

Manitoba was the first new province of Confederation, joining in 1870. This was some three years after the provinces of Canada (which became Ontario and Quebec), Nova Scotia, and New Brunswick signed the British North America Act. Although the older provinces had retained control over both provincial lands and local public works, the situation was different in Manitoba because rapid settlement in the prairie region was deemed essential to the establishment of a transcontinental nation.[7] By assuming authority over the lands of the prairie region, both in the province of Manitoba and in the territories farther west, the dominion government put itself in a position to manage the initial aspects of regional settlement. Yet though the Canadian government was very concerned with getting people onto the land, it was far less worried about how settlers fared once they were there. Although Ottawa maintained control over the lands of the province, the Manitoba government, like the governments of the other provinces, was accorded responsibility for local infrastructure. By Confederation in 1867, the eastern provinces already had extensive public works systems, including roads, railways, and public buildings. Manitoba had far less in the way of public works development when it joined Confederation some three years later, and establishing the necessary transportation infrastructure was a key challenge for the new provincial government.

Although Manitoba lacked an extensive road system, it was hardly untravelled. Before 1870, routes through the region generally reflected the character of the landscape. Overland travellers favoured well-used routes along the natural levees of the Red and Assiniboine Rivers.[8] Farther from the riverbanks, early trails followed ridges and bypassed low areas.[9] Efficient travel usually meant taking the driest – rather than the most direct – line between two places. These routes were, in the judgment of Department of Public Works employee J.A. Macdonell, "the best natural roads" of the region and became part of the shared resources of the community. But avoiding wet areas was not always possible. Macdonell noted that, if the natural roads of the region were obstructed for any reason, the traveller would have "no alternative but to plunge too frequently through some times dangerous and almost impassible swamps and mud-holes."[10] Engineer H.A. Bowman, a long-time employee of the Department of Public Works, remembered, in the early period, "women going [to] or returning from market who had to hold their clothes up round their waist when wading the muskegs."[11] And even the natural roads of the region could present challenges. In an 1880 letter to a loved one outside the province, Lucinda Westover, wife of Asa Westover, who had taken a temporary position as farm manager in south-central Manitoba, described travel along prairie trails. Mrs. Westover was impressed by the trails that seem to "drive anywhere in any direction through the grass." But still she worried about the risk of accidentally becoming bogged down in a wet area, which like Bowman, she called a muskeg. In her view, familiarity with the landscape was necessary to determine "by the appearance of the grass what the bottom is, or whether there is any bottom," to any unavoidable "muskeg."[12] For a provincial government keen to encourage settlement, such risk and inconvenience were unacceptable. From an early period, it was clear that facilitating travel and transport through the region meant undertaking drainage along highways. Because of this, some of Manitoba's first drains were built along roads.

The need for road drainage increased exponentially as road locations came to be determined less by local conditions and more by the township survey. One feature that distinguished the Canadian survey system from the American system on which it was modelled was the inclusion of what were called road allowances.[13] In southern Manitoba, these were strips of 90 feet (27.4 metres) marked out between all townships and sections. Although construction and maintenance of local roads were provincial responsibilities, the location of many new major roads was determined by the Dominion Land Survey. This method of land division helped to ensure that the necessary land would be available, but it did not guarantee that the road allowance would be environmentally suitable for road building. In the 1940s, engineer F.E. Umphrey, a long-time employee of the Department of Public Works, mused in a letter to an associate that, if environmental conditions had been the primary determinant in the province's early road-building efforts, "we would probably find many of our roads leaving the right angled section line location for angular location following the sand ridges and higher land, and our drains following the lower lands where they should be, if required at all."[14] As it was, with road locations defined without regard to local land conditions, new highways often required extensive drainage.

Both the dominion and the provincial governments wanted to encourage immigration, but their interests were not identical. Whereas the dominion government derived benefit from the settlement of lands, a process that was relatively easy to administer once the land had been surveyed, the provincial government was responsible for the public works such as roads and drains that were necessary to make agricultural settlement viable within the survey grid laid over the irregular and dynamic landscape of Manitoba's low areas. Manitoba officials realized early on that increasing settlement "was a most serious problem for the provincial treasury" because of the increased demand for public works that would inevitably result.[15] In the late 1940s, engineer and former deputy minister of the Department of Public Works M.A. Lyons, in describing the settlement history of an area south of Lake Manitoba, recalled how, "almost immediately" after the dominion government opened these lands to settlers, "the provincial government was besieged with requests for drainage." Public works construction varied in difficulty and expense depending on environmental conditions, Lyons noted. Had the dominion government done more to dissuade settlers from occupying unsuitable lands, the work of the province might have been substantially reduced. Consequently, Lyons thought that dominion land settlement policy in effect had left "the provincial government holding the bag."[16]

Early drainage was largely aimed at improving the transportation infrastructure not only because of the need to open up routes through the province but also because flooding away from designated road allowances was not yet perceived as a major problem. The settlement grid did not itself accommodate environmental irregularity, but early settlers working within it had sufficient choice to enable them to avoid unsuitable land. Drier sites did indeed fill up more quickly.[17] However, while locating in a swamp was clearly a disaster, establishing a home alongside wet areas presented certain advantages. Some regions even marketed themselves by emphasizing advantages derived from their comparatively wetter character, such as good hunting and easy access to hay.[18] Those who followed the early arrivals made decisions about where to settle based on the human as well as the physical geography of the province, considering proximity to family and friends or those of similar cultural origin in addition to land conditions.[19] The result was a settlement pattern that curved around wet areas but without much of a buffer between drier lands newcomers envisioned as productive farms and wetter lands they perceived as comparatively useless.

Unfortunately for those who settled close to wet areas, the wet prairie landscape was characterized not only by spatial variability (the wet-dry pattern noted by Hind) but also by temporal variability (change in the pattern over time). So lands selected as dry did not always stay dry, and locations that provided ready access to wetland resources often proved vulnerable to flooding.[20] Unexpected flooding could result from various factors. High rates of precipitation over a number of years could cause local watering holes to expand over neighbouring farmlands. A quick spring thaw or a few weeks of heavy rain could lead to pooling, especially in areas of relatively impermeable clay-based soils. Some farmers settled lands that were more often wet than dry, having had the misfortune to make homestead entry in a dry interlude or having made settlement decisions based on factors other than environmental conditions. In 1929, after substantial government investment in drainage infrastructure, H.A. Bowman, who had become chief engineer of the Department of Public Works, was asked to summarize the history of drainage in the province. In his opinion, settlers had occupied land that "should never have been put under the plough." Settlers broke the land in dry cycles, and, "when the wet cycles recur, there is an immediate cry for help."[21] But a few decades earlier, Bowman himself made comments making it clear that, despite the far greater role of non-anthropogenic environmental variation, the government was not entirely blameless. In explaining the work of the department, Bowman noted that, given the pioneering character of many road-building operations, "it is not possible in many districts to take the water to an outlet, which in numerous cases is many miles away."[22] By directing water away without paying attention to where it was going, well-intentioned drainers built drains that could serve to create or exacerbate flooding on nearby lands, perhaps even rendering homesteads less suitable for agriculture. For a number of reasons, even settlers who thought they had chosen their lands wisely could end up in areas vulnerable to flooding.

. In May 1896, settler G.S. Howard wrote to the minister of public works on behalf of his neighbours in the Cromwell district, located east of the Red River. The letter and accompanying petition provide a sense of what it was like to live in a frequently inundated area: "It is injurious to our health to go about from day to day wet footed. Anywhere we want to go we have to walk through water to get there. Our children cannot go to school half the time they should go and they are [losing] their education in consequence." In the minds of these settlers at least, local flooding was more than an inconvenience. Living in the wet prairie posed serious difficulties for settlers hoping to achieve a reasonably pleasant life for themselves and improved circumstances for their children. And when agricultural production was brought into the picture, the situation seemed only more desperate. Since settlers were unable to "grow anything when the land is so wet," starvation seemed to be a real risk for many. Letters such as Howard's made it clear that good surface drainage was necessary not only to ensure the establishment of a reliable transportation infrastructure but also to bridge the gulf between the agricultural aspirations of newcomers and the environmental realities of southern Manitoba. As Howard put it, "if the Government want settlers to come to Manitoba, they ought to make the place fit for them to live in."[23]

The Manitoba government would try to do just that. Good surface drainage has been described as an interconnected network of appropriately sized channels, tree-like in how the smallest collector drains dump into larger channels that eventually lead to substantial trunk drains.[24] To replicate the tree-like pattern characteristic of effective natural drainage, artificial drainage required a coordinated infrastructure, with logical gradation among channel sizes and suitable outlets in substantial natural waterways. If these were not achieved, drainage projects could worsen existing surface water problems or even create new ones. Secure funding was necessary to support the careful project design and execution essential for the construction of good drains. In the early years, the government undertook drainage in an unsystematic fashion, with the public works department throwing money and resources at problem areas.[25] But it was ultimately government authority that would make possible the group finance and coordinated construction necessary for more successful drainage. Smaller, on-farm drainage likely took place through individual effort. Some drainage might have been achieved through neighbourly cooperation, though early on opportunities for this were limited in many still sparsely settled areas. Large-scale land change in southern Manitoba was achieved through government management, as institutionalized administration offered ways of coordinating the work and managing the finances.

In 1880, Manitoba passed its first drainage legislation. Under The Drainage Act, the province would undertake to drain nine large wetland complexes through the construction of nearly 200 miles of drains.[26] The wetlands to be drained were the St. Andrews Marsh, the Seine River Marsh, the Springfield Marsh, the Boyne River Marsh, the Westbourne Marsh,

FIGURE 2    Men and horses digging a drain. This channel was located west of Lake Manitoba. The implements pictured are relatively simple, the sort that many farmers would have used on their farms.
*Source:* Archives of Manitoba, Drainage 26, N 23022.

the Big Grass Marsh, the Woodlands Marsh, the Tobacco Creek Marsh, and the marshes southwest of Rat River in and around Provencher.[27] As explained in the minutes of an earlier Executive Council meeting, the legislation was designed to address a particularly worrisome situation. Because of these large wet areas, "immigrants were either deterred from entering the province, or were forced to pass through it and settle on the drier plains beyond."[28] This was the province identifying the wet areas that were the most blatant contradictions to the agricultural landscape new-comers expected to find and committing to extending its administrative and public works capacity to address the problem.

This was also the province reacting to certain realities in its human geography. Although the settlement taking place created real challenges for the province in terms of infrastructure construction, Manitoba was not growing at a remarkable rate, and that would not change any time soon. Despite a significant boom in the land market in the early 1880s, recovery from the subsequent bust was slow. In 1883, 1,831,982 acres of land were alienated, down over 800,000 acres from the previous year's figure. In 1884, only 1,110,512 acres were taken up. And in 1885, settlers laid claim to 481,814 acres, a shockingly low number.[29] Rates of growth would not fully recover until 1897.[30] The 1880 drainage legislation addressed the flooding that was already perceived as a serious barrier to provincial development, seeking to ensure no one was discouraged from taking up land in the province. It confirmed the provincial government's important role in early drainage, in the context of real concern that continued flooding would impede settlement, at a time when the future of the province looked far from assured.

Despite its apparently public-spirited aims, the 1880 legislation caused concern among some Manitobans. With regard to these and other undertakings, there were suspicions of corruption in drainage contracting. Indeed, at least some of the public money invested in drainage was distributed in ways that served political ends. For example, in August 1892, Robert Wemyss wrote to Premier Thomas Greenway to explain how the Council of the Rural Municipality of Lakeview regularly expended public money (likely including grants from the province) on drainage projects "which while benefitting one or two people is of no use to the general body of settlers." Municipal councillors and their allies stood to gain from these undertakings as contracts let to political friends (rather than to the lowest bidder) shored up support. Wemyss perceived a relation between municipal and provincial politics and warned Premier Greenway, a Liberal, that the local municipal council was distributing drainage contracts in such a way as to favour the opposition Conservatives.[31] Like other public works undertakings in Manitoba and elsewhere, drainage became a political instrument, a means by which those in power could cultivate support.

There were other ways that party politics and drain construction affected each other. Not content to allow the provincial or municipal government to determine which projects should be undertaken, some settlers fashioned political leverage out of their experiences of flooding. They applied it to the government of the day, sometimes with seemingly little regard for political loyalties. One particularly telling example was a letter from farmer Oswald Berire to Premier Greenway. Although he had "voted grit for 40 years," Berire threatened that, if a government ditch in his area were not finished that fall, "I will turn my coat and there is some more of my neighbours like me."[32] While some settlers, such as Berire, threatened the government with voter dissatisfaction, others simply underlined that authorities had a responsibility to them, convinced that the provincial government was at least partially at fault for flooding. Some went so far as to entirely blame the authorities for their situation, claiming it was "owing to the energetic immigration policy of our Provincial government" they had taken up land during dry periods that was flooded in wet periods, apparently oblivious to the fact that the federal government, not the provincial, had solicited immigration.[33] Beyond surface water, troublesome enough on its own, provincial governments were also confronted with the task of managing settler discontent.

For the provincial government, the situation was rendered still more challenging by objections from settlers to drainage projects. In 1887, farmer Thomas Usher wrote to Department of Public Works employee Wilson to complain about a newly constructed drain that cut across his farm in two places, with his "grain fields being on both sides of the ditch." In the absence of multiple bridges, Usher foresaw the ditch would be the cause of "a great deal of inconvenience." His dismay was exacerbated by the fact that he did not himself stand to gain from drainage.[34] Drainage ditches could be problematic because of their size as well as their location. As settler J.O. Smith explained to a Department of Public Works employee in 1886, the narrow ditch made by the department posed a risk to local cattle. As Smith put it, "if you like your [beef] I think you will be getting some on your market that will die in the dit[c]h with their feet the [w]rong way up." Smith wished that the ditch had been built "wide enough for an animal to right itself if it should by accident get in on its back."[35] The Department of Public Works received complaints about those ditches that had been built as well as those that had not, redoubling the complexity of the wet prairie landscape.

The government of Manitoba became involved in land drainage for a number of reasons. Most important was the discrepancy between the regular, grid-based settlement supported by the Dominion Lands Act and the irregular, variable pattern of surface water across many areas of the province. With responsibility for local public works, the province had the difficult task of reconciling the two. Beginning with road drainage and progressing to agricultural drainage, the Manitoba government sought to provide the necessary supports to newcomers. This was in keeping with the liberal practice of providing an infrastructure to support capitalist development. At this early period, Manitoba's future was far from assured, and the province took up drainage with a vigour inspired by real fear that inaction might retard development. As provincial employees and contractors were hard at work in the wet regions of the province, engaged in the physical process of environmental transformation, they themselves were evidence of liberalism at work on the ground.

Newcomers arrived in Manitoba expecting to find a ready agricultural landscape. Over a period of years, many realized their lands were subject

FIGURE 3    Boating in a drainage ditch near Woodside, Manitoba. This photograph is
rare evidence of recreational engagement with Manitoba's drainage infrastructure. Note
that both figures in the boat seem to be women, whereas workers on drainage projects
typically were men.
*Source:* Archives of Manitoba, Jessop 178, Drainage Ditch, Glen Farm, Woodside, N 3211.

to surface water flooding, and the problem went beyond what they could
cope with on their own. Many turned to the provincial government for
assistance, writing letters that record their hopes for the land as well as
their disappointments with the current situation. Some Manitobans seem
to have written in the expectation of receiving government assistance. For
these individuals, the liberal ideology by which states would assist with
infrastructure projects might well have been part of the baggage they
brought with them to the province. Other Manitobans seem to have written out of sheer desperation, believing government assistance was all that
stood between them and utter ruin. The pattern in both cases, however,
was that of residents appealing to the provincial state.

But if settlers shared an expectation the government would assist with
drainage, they did not share a common opinion about what should be
done and how the government should do it. Drainage might have been
consistent with liberalism, but it did not remain in the airy realm of intellectual abstraction. Drainage quickly became imbedded in the day-to-day

ins and outs of provincial politics and community disputes. The dynamic between politics and drainage was significant, with Manitobans both expressing outrage at perceived government corruption and leveraging government funds through threat of voter discontent. The government undertook drainage work under a variety of pressures, only some of them environmental. And the matter was only more fraught in cases where settlers perceived that government drainage had somehow worsened their situation.

## MUNICIPAL GOVERNMENTS, DRAINAGE DISTRICTS, AND ETHNIC IDENTITY

The Manitoba government's liberal inclination to undertake drainage was bolstered by fear that only ready assistance would retain disheartened settlers. Yet government finances were as precarious as the provincial population, and economic pressures limited the drainage works the province could undertake. Growing demand for agricultural drainage combined with continued provincial inability to satisfy all requests for assistance meant that those who sought drainage often joined the push for municipal incorporation. Municipalities were corporations created by provinces for the purposes of local administration.[36] Their powers included the capacity to carry debts, to levy and collect assessments, and to undertake public works of local significance. As municipal governments levied local taxes, locally desired projects could go ahead without provincial support. A municipality could also operate as a lobby to leverage financial assistance from the Department of Public Works. Any funds raised would be expended on locally significant projects, according to the will of the municipal council. The coordinated administration and group finance that earlier had been possible only through the provincial government were available within local communities once municipal government was established.

Municipal boundaries in Manitoba were irregular, reflecting local patterns of settlement and local opinions about when to incorporate. Once established, municipal governments were very good at certain things. For instance, they provided an effective way to manage road construction. Ideally, the road system would stretch across the municipality, with bridges, culverts, and roadside drains allowing the grid-based system to trump natural drainage patterns. As individual municipalities worked within the settlement grid that had been extended across the province and beyond into the Northwest Territories (now the provinces of Saskatchewan and

Alberta), the line where one municipality stopped and another started was relatively inconsequential. Even if one municipality set to work years before another, it was likely that improved roads would eventually line up. And in the interim, a finished road would cause little more than envy in the neighbouring municipality. Over time, and despite the difficulties that attended intensive construction projects in a still remote province, municipal road building would add up to an effective provincial road system.

The matter of agricultural drainage was less straightforward. There was no grand plan bearing on drain layout in the way that the road allowances of the Dominion Land Survey guided road locations. Neighbouring municipalities, even when all were interested in undertaking drainage, could come to radically different ideas of the work necessary.[37] A drain that ended at a municipal border could change water patterns in a manner detrimental to nearby residents. Although increased reliance on engineering professionals at the planning stage served over time to minimize instances of this problem, the matter remained a concern.[38] And even when the need for inter-municipal coordination was recognized, there were difficulties in negotiating terms of cooperation that suited all parties. Municipal drainage could compound the problem of unsystematic drainage, serving to displace or even worsen surface water flooding rather than to resolve it. The problems stemming from piecemeal drainage became more apparent as rates of settlement increased in the 1890s, because greater population density made it more likely that irresponsible drainage would contribute to the flooding of cultivated land. In Manitoba as elsewhere, accumulating private complaints increased the pressure to find general public solutions to water management problems.[39]

Divisions of interest within municipalities were also troublesome. Everyone needed roads, even if construction priorities were the subject of hot debate. The patchy nature of surface water conditions meant that, even in the wettest municipalities, not everyone would benefit from drains. As a result, not everyone favoured the expenditure of municipal resources on drain construction. The problem was only expanded at the provincial scale since the consolidated revenues of the province were expended on resolving a problem that did not afflict all regions equally. At both levels of government, there were grounds for the complaint that money that should benefit many (all in the municipality or all in the province) was only benefiting a few (those subject to flooding). The expenditure of public money on drainage that largely benefited private lands was problematic.

And as larger projects were contemplated, the current funding arrangements came to seem increasingly unsatisfactory.

The difficulty of coordination across municipal boundaries and the problem of financing more substantial undertakings comprised a large part of the impetus for the creation of drainage districts. The drainage district idea was not unique to Manitoba. By the end of the nineteenth century, such entities were standard in large-scale, government-assisted land drainage projects. Between 1857 and 1932, drainage district legislation was enacted in thirty-five American states, including those with environments comparable to Manitoba such as Minnesota and North Dakota as well as those with dramatically different landscapes such as Florida and California.[40] Like other forms of special district government such as irrigation districts or school districts, drainage districts were administrative entities created to facilitate the undertaking of a specific task. They amounted to a targeted solution for the problem of flooding, aimed at those areas of Manitoba in which surface water problems interfered with the private property landscape of the Dominion Land Survey.

Manitoba's 1895 Land Drainage Act contained the legislative provisions for the creation of drainage districts. These districts represented another set of lines across the geography of the province, lines that related to surface water conditions in a way the township survey and the municipal system did not. Defined geographically by perceived flooding problems, drainage districts made possible a more direct relation between those who benefited from drainage and those who paid for it, through the creation of a taxable entity that approximated the flooded area. Through the sale of debentures on behalf of districts, the provincial government would front the money for large-scale drainage, with residents of the drainage district repaying the loan over a period of decades. The expectation was that the improved agricultural productivity of drained land would more than compensate for the expense of construction. Furthermore, in the abstract at least, drainage district residents were unlikely to resent the expense since they were simply splitting the bill to solve a shared problem. Dividing the costs among those who would benefit was comparable to a farmer paying for improvements to a homestead. Ultimately, drainage districts seemed to address the problem of public money and private lands, ensuring that people without problems of flooding were not underwriting the costs of constructing drainage works for those at risk of flood.

One of the factors common to drainage throughout the province was the use of surface ditches. Drainage districts abutted the major lakes and

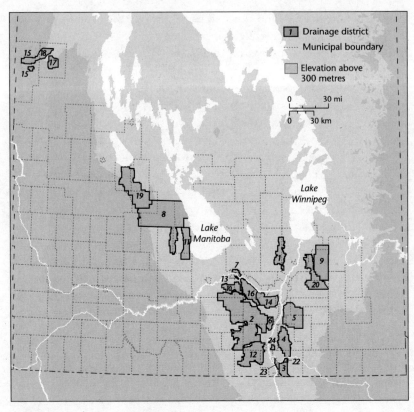

FIGURE 4   Drainage districts in Manitoba, 1933
*Sources:* Adapted from Archives of Manitoba, Manitoba Department of Mines and Natural Resources, Surveys Branch, Map of Manitoba, Southern Portion, 1933; the Drainage Districts, map,
in J.H. Ellis, *The Soils of Manitoba* (Winnipeg, Manitoba: Economic Survey Board, 1938), 29.

rivers of the province and either dumped directly into them or, more often,
into smaller tributaries. The logistics of draining Manitoba were relatively
simple. A substantial amount of work was done by horse and plow, though
large ditch-digging machines were also employed. In Drainage District
No. 1, for instance, two dredges were built on location. They started digging at the high end of one of the major drains and floated downstream
in their own ditch as they worked toward the outlet. The massive, steam-
powered dredges were slow, but they completed a significant amount of
excavation in early twentieth-century Manitoba.[41] Moreover, these impos-
ing machines lent a sense of grandeur to the enterprise, confirming the
association between drainage and progress.

FIGURE 5    Posing with ditch-digging machinery. The machine, the car, and the well-dressed visitors seem to be presented as symbols of progress and modernity. *Source:* Archives of Manitoba, Foote 208, 1914 Drag Line (Neg 1808).

These massive drainage machines operated amid Manitoba's developing human geography. Over the settlement period, newcomers from Ontario and Britain came to predominate in some of Manitoba's most fertile lands, particularly in the Southwest Uplands, which were not particularly vulnerable to surface water flooding. This arrangement was partly a matter of timing, as British Canadians claimed much of the best land at an early date.[42] Under the influence of Minister of the Interior Clifford Sifton, who favoured the immigration of groups he understood to be experienced farmers, a later large-scale influx emphasized central and eastern Europeans. Many of these newcomers settled in the less productive lands along the margins of the Precambrian Shield in eastern or interlake Manitoba.[43] Ultimately, neither the rich southwest nor the rocky margin was the main target for drainage. Beyond the Manitoba Escarpment, there was little need for it. Indeed, this area was part of the prairie tabletop that ran west to the Rocky Mountains, and it suffered more frequently from water scarcity than overabundance. Areas with the inadequate soil and plentiful rocks characteristic of the Precambrian Shield were largely excluded from drainage, as even if entirely successful the removal of surface water was insufficient to create conditions conducive to prosperous agriculture. Best suited for drainage were those areas with rich soil but excess water, which were largely confined to the Red River valley and lakeshore regions of the province. The human geography of these areas was diverse. Through a survey of 157 farms in the Red River valley published in 1936, sociologist R.W. Murchie found that "all of the major European groups were well represented, indicating that the population in this area is distinctly mixed, ethnically."[44] This sample was suggestive of conditions in Manitoba's wet prairie throughout the settlement period, as people of different back-grounds sought the common goal of agricultural success.

Since the 1960s, ethnic identities have framed how many scholars have understood the history of Manitoba.[45] Given the cultural diversity evident in even a relatively small area such as the Red River valley, such an approach clearly had much to offer. Perhaps most importantly, this approach under-lined the plurality of the Canadian experience, emphasizing J.M.S. Careless's "limited identities" instead of an overarching national narrative.[46] Recent scholarship on ethnicity has revealed the diversity and dissonance within any one ethnic group as well as the shared influences bearing on all contemporary groups within Canada and around the world.[47] Emphasizing the ecological commons of the wet prairie adds another layer to our understanding of the lives of newcomers to Manitoba, indicating some of the environmental influences that, along with cultural forces, bore on their lives. This perspective underlines the continuities that span the distinctions between ethnic groups while linking these continuities to the specific geographical context of the wet prairie.

Although the boundaries of Manitoba's drainage districts corresponded to the physical geography of the flood problem, they diverged from the province's human geography, spanning meaningful political and cultural boundaries. Among the most culturally distinct of Manitoba's municipal-ities were those that grew out of the province's Mennonite reserves. Mennonites trace their lineage to Menno Simons, a sixteenth-century re-ligious leader. They are traditionally committed to certain principles, in-cluding adult baptism, non-violence, and the primacy of religion in community life. Mennonite cultural heritage includes a history of difficult relations with civil authorities due to unwillingness to obey any secular authority not in accord with the tenets of the religion.[48] Between 1874 and 1880, nearly 7,000 Mennonites emigrated from Russia to Canada.[49] Federal government efforts to encourage prairie settlement helped to spur this influx. In 1876, Ottawa amended the Dominion Lands Act to allow for what was called "group settlement."[50] The amendment accommodated groups wishing to establish land management systems that deviated from the prairie standard. Under this amendment, two Mennonite reserves were eventually established in Manitoba, one to the east and one to the west of the Red River.

Mennonites were drawn to Manitoba in part by the possibility of preserving their distinct lifeways. As historical geographer John Warkentin explained, "this was not a migration of individualistic pioneers,"[51] and their land use practices reflected this. On both the east and the west reserves, agricultural villages were formed, with families living in proximity to each other. A system of village governance was established, through which decisions bearing on the entire community were made.[52] The land surrounding the village was divided into a few large fields according to the purpose to which it was most suited, such as agricultural crops or hay growing. Each field was then divided into strips for use by individual village residents. Historian Royden Loewen has explained that along with this distinctive settlement pattern went Mennonite self-perception that emphasized continuity with those who shared their faith and their lifestyle and distinctiveness from those who did not.[53]

The Mennonite land management system allowed the community a greater measure of flexibility in accommodating local environmental conditions, including surface water patterns, than was available to the average homesteader.[54] Still, there was some immediate concern among the Mennonite community over waterlogging and other land quality issues on the east reserve.[55] The greater productivity of the west reserve lands was soon abundantly clear, and some Mennonites who had found it too difficult to make a living on the east reserve moved across the river.[56] But the west reserve was hardly immune to surface water problems. Indeed, the area had been available for Mennonites in part because many early settlers had avoided it due to the frequency of flooding.[57] Warkentin has argued that the "steady cooperation" that characterized the Mennonite approach to agriculture made possible large-scale projects that even the "periodic cooperative work bees of Ontario Canadian settlers" could not match, suggesting that Mennonite communities might have been more capable of the coordinated work necessary for successful drainage.[58] Regina Neufeld's history of the east reserve village of Schantzenberg includes a description of an incident that bears this out. Neufeld tells of her great-grandfather's participation in a group ditch-digging effort, which resulted in a hand-dug drain some six to eight miles long.[59] But despite ready co-operation and hard work, persistent problems with flooding hindered agricultural prosperity on Mennonite lands both east and west of the Red River. The problem, at least in part, was that Mennonite communities were embedded in the larger ecological commons of the wet prairie.

Newcomers to Manitoba located themselves within the grid-based settlement pattern established by the federal government, but flowing water did not respect the lines on the land. Just as water crossed property lines, so too it spanned municipal boundaries, some of which corresponded to transitions between cultural groups. Ultimately, agricultural flooding took place with little regard for the human geography of the region. As water ran from one local jurisdiction or ethnic concentration to another, it created a shared concern among Manitobans, some of whom might otherwise have had little in common. Importantly, the drainage infrastructure created to address the problem of flooding also spanned divisions between homesteads, municipalities, and cultural groups. Artificial drainage was intended to manage the ecological commons of surface water, to make it possible to farm successfully within the settlement grid. But to carry water swiftly and smoothly, drainage channels had to extend across lines of property and community in ways that mirrored the underlying natural drainage system. They became the constructed incarnation of the ecological commons, the product of an encounter between the liberal government of Manitoba and the wet prairie landscape.

The human consequences of flooding and drainage were especially significant for groups that placed particular value on their cultural distinctiveness, such as the Mennonites. When the provincial government passed stronger municipal legislation in 1880, and despite the objections of some traditional Mennonites who wanted to remain aloof from the non-Mennonite community, the Rural Municipality of Rhineland was formed out of the Mennonite West Reserve.[60] Despite whatever concerns the Mennonite community might have had about closer ties with a civil government, the newly institutionalized relationship between the municipality and the province presented certain advantages. For instance, in light of persistent flooding in the 1880s and 1890s, the Rhineland municipal council appealed for assistance with drainage projects and received a significant grant.[61] As in other municipalities throughout the province, there were real financial benefits to Rhineland's formalized relationship with the province.

However, as historian Gerhard Ens has observed, early drainage projects in the Rhineland area served only to confirm "that piecemeal drainage and dredging [were] not going to alleviate the problem."[62] Further action was necessary. In July 1894, C. Hiebert, secretary-treasurer for the Rural Municipality of Rhineland, wrote to the minister of public works seeking advice on a drainage problem. Rhineland was keen to press ahead with further drainage but had been unable to negotiate an arrangement with the neighbouring, largely non-Mennonite, Rural Municipality of Montcalm, through which Rhineland's excess water would flow before reaching a waterway that led to the Red River. Personal negotiations between Rhineland Reeve Heppner and the appropriate Montcalm officials had failed to produce results. Hiebert went on to explain that the ditch in question was one among the many ditches that would be necessary if the Rhineland municipal council was to succeed in giving "every farm or farmer a chance to make a living on his property."[63] In 1899, the Rhineland municipal council petitioned the government to bring three of its townships under the recently passed Land Drainage Act.[64] A drainage district that spanned municipal boundaries would provide a way of circumventing differences of opinion between municipalities and allowing necessary drainage work to proceed. It would create a common concern for what was, according to the 1901 census, the 96 percent German (likely largely Mennonite) population of Rhineland and the 71 percent French population of Montcalm.[65] And it would provide an administrative infrastructure appropriate to the ecological commons that encompassed both municipalities.

Mennonite commitment to drainage is further documented in a 12 March 1903 article in the Manitoba newspaper *Morning Telegram*. The sympathetic reporter described a political meeting held in Emerson, a town slightly north of the US border and immediately east of the Red River. Premier Roblin and Provincial Secretary McFadden spoke to a packed house at the Emerson Town Hall. The audience included "considerable numbers of Germans," many of whom were likely German-speaking Mennonites. As described in the article, the meeting focused in large measure on the issue of drainage, with the speakers addressing the political and environmental repercussions of particular drainage decisions.[66] Given their history of resistance to civil authority, at least when it conflicted with religious principle, some Mennonites might have felt uncomfortable with the singing of "God Save the King" that concluded the meeting.[67] But they had no trouble recognizing their stake in the drainage projects of the provincial government. Residents of the wet prairie shared an economic interest in the government-sponsored drainage projects that were necessary for the production and protection of agricultural land. This shared interest amounted to a meeting ground between culturally distinct communities.

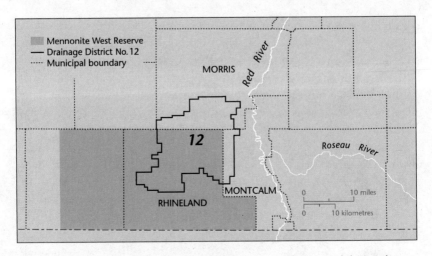

FIGURE 6   The Mennonite West Reserve, Drainage District No. 12, and the Rural
Municipalities of Morris, Rhineland, and Montcalm
*Sources:* Adapted from Archives of Manitoba, Manitoba Department of Mines and Natural Resources, Surveys Branch, Map of Manitoba, Southern Portion, 1933; the Drainage Districts, map,
in J.H. Ellis, *The Soils of Manitoba* (Winnipeg, Manitoba: Economic Survey Board, 1938), 29.

In April 1903, a report of a committee of Manitoba's Executive Council
recommended that Drainage District No. 12 be created in an area that
included significant Mennonite lands.[68] The drainage district would include
parts of the neighbouring Rural Municipalities of Montcalm and Morris.
No adequately supported petitions against drainage were received (as required by provincial legislation, petitions would be disregarded unless they
had the support of landowners representing more than half the value of
the affected land), so the district was proclaimed. The scope of the project
was enlarged as years passed, and construction continued through 1907.
Even so, the desired drainage was not achieved. In the late 1910s, Rhineland
and other municipalities, including Montcalm and Morris, formed the
Red River Valley Drainage and Improvement Association to lobby the
provincial government more effectively. A community history of the Rural
Municipality of Morris provided a list of participating municipalities, and
Table 2 illustrates the national origins of their residents as per the 1921
census. [69]

The ethnic makeup of the area represented by the Red River Valley
Drainage and Improvement Association is further revealed by considering
the percentage of the participating municipalities' total population that
was represented by each of the most prominent nations of origin. The

TABLE 2

**National origins of residents of rural municipalities involved in the Red River Valley Drainage and Improvement Association according to the 1921 census**

| Municipality | National origin Most to least prominent (%) | | | | | | |
|---|---|---|---|---|---|---|---|
| Morris | German 31.3 | Russian 27.3 | French 14.5 | English 10.4 | Scottish 6.1 | Irish 4.3 | Dutch 0.3 |
| Dufferin | English 36.6 | Scottish 26.1 | Irish 18.7 | French 5.6 | German 2.7 | Dutch 0.9 | Russian 0.3 |
| Macdonald | English 20.5 | German 16.1 | Scottish 15.8 | French 14.6 | Irish 12.6 | Russian 2.1 | Dutch 1.1 |
| Montcalm | French 73.2 | Dutch 7.1 | English 5.1 | Scottish 4.3 | German 4.0 | Irish 3.4 | Russian 1.2 |
| Rhineland | Dutch 88.6 | Russian 5.5 | German 4.3 | English 0.5 | Scottish 0.2 | Irish 0.1 | French 0.1 |
| Roland | English 29.1 | Irish 23.6 | Scottish 20.9 | Russian 17.8 | German 2.6 | Dutch 2.0 | French 0.4 |
| Thompson | English 38.4 | Scottish 26.2 | Irish 21.1 | Russian 3.2 | German 2.3 | French 1.7 | Dutch 1.3 |

*Notes:* Percentages do not add up to 100 percent as ethnicities not within the top three for any of the listed municipalities were omitted from the table. Mennonites were identified as Dutch rather than German (as in earlier censuses) due to lingering wartime prejudices. Scottish has been substituted for Scotch.

*Source:* Adapted from Statistics Canada, Table 27, "Population Classified According to Principal Origin of the People by Counties and Their Subdivision, 1921," 1921 census.

largest group was made up of those indicating their national origin as Dutch (likely largely Mennonite) at 26.6 percent, followed by English at 14.8 percent, French at 13 percent, Scottish at 10.4 percent, German at 9.7 percent, and Irish and Russian both at 8.6 percent.[70] With the Dutch excluded, the remaining six groups fall within 6.2 percent of each other, reflecting a relatively similar numerical representation of these nations of origin within the area included in the Red River Valley Drainage and Improvement Association. If the percentages for the English, Scottish, and Irish are summed to reflect a possible affinity among those whom the 1921 census grouped as the British races, then the combined value is fairly similar to the percentage of people of Dutch national origin: 33.8 percent British compared with 29.6 percent Dutch. The remaining three groups (German, Russian, and French) together also amount to roughly a third

of the area's population. Although the population in the area represented
by the Red River Valley Drainage and Improvement Association does not
necessarily reflect participation in the association, it seems clear this organization worked on behalf of a region of the province united less by
nation of origin than by shared environmental experiences. Common
grievances brought these distinct communities together, as an organization
positioned to work on their behalf pushed the provincial government for
further action on agricultural flooding.

Geographer W.J. Carlyle found that Mennonite farmers made crop
choices according to environmental conditions rather than ethnic inclina-
tions.[71] Historian Royden Loewen noted the evolution of Mennonite
agricultural practices in response to environmental conditions in Manitoba's wet prairie.[72] Studying drainage takes the matter further, illustrating
how shared environmental conditions created common experiences as well
as similar agricultural landscapes. Because of the ecological commons,
successful drainage meant cooperation across cultural and municipal
boundaries. The Mennonite experience, though particular to that culture,
might suggest how environmental circumstances and cultural factors
interacted for other groups in the ethnically mixed area of Manitoba's wet
prairie. A population divided by country or culture of origin was nevertheless united by similar environmental problems and cooperative infrastructure solutions. Furthermore, at least some residents looked beyond cultural
differences to see the potential in working together to express dissatisfaction
with the current situation and to advocate for more effective surface water
management.

To live within a drainage district was to occupy a particular relation to
the local landscape and to a drainage system created by the provincial state.
Drainage served to involve Manitobans from different cultures in a large-
scale, long-term environmental project. Surface water flowed through
Manitoba's cultural groups as distinct communities became embroiled in
drainage projects that created common concerns among different groups.
Insofar as drainage depended on the intervention and assistance of the
provincial government, Manitobans of various backgrounds also came to
share a parallel orientation toward the provincial state. And as the solution
of drainage proved less than effective in some instances, this orientation
was characterized by anger and protest as well as cooperation.

From all of this emerged experiences common among many Manitobans
that were generated partly in relation to the wet prairie environment. In
analyzing the history of the province, these shared circumstances are worth
considering along with distinguishing factors such as religion, language,

FIGURE 7   A dredge in a flooded area. Few historical photographs of drainage machinery include wet terrain, perhaps because drainage work was more effectively executed during dry periods.
*Source:* Archives of Manitoba, Public Works/Drainage 8, C75/2, n.d., Cross Walking Dredge on Brown Drain, H.A. Bowman – Chief Engineer, Western Drainage Company.

and ethnicity. The wet prairie was the catalyst to the development of subject positions that cut across the cultural identities that figured large in the historical experiences of many Manitobans and that have framed how many scholars have understood the history of the province. The ecological commons of the wet prairie and the drainage channels that tracked through it were the environmental basis for local communities of interest defined in relation to surface water patterns.

## THE ABORIGINAL EXPERIENCE

The situation was different for Aboriginal groups. The cultural conduits of drains did not extend across the racialized boundary between Aboriginal and non-Aboriginal lands. The comparative similarity of Mennonite and non-Mennonite experiences with drainage – the common layer of experience derived from the difficulties of flooding and the challenges inherent to cooperating with neighbours and governments on solutions – is emphasized through contrast with the distinctiveness of the Aboriginal story. With respect to flooding, Aboriginal reserves were no better positioned than Mennonite reserves. But though the two groups occupied a similar position in relation to surface water, their experiences of the process of drainage were entirely different.

In the late eighteenth century, the Ojibwa moved into what became southern Manitoba, establishing themselves in territory that had recently been abandoned by the Assiniboine and the Cree.[73] They came from what is now northwestern Ontario, an area that included many wetlands.[74] Ojibwa groups settled in various locations, with some heading to remote areas and some establishing strong ties to Netley Marsh, a lacustrine wetland complex in the Red River delta immediately south of Lake Winnipeg. George Van Der Goes Ladd, a United Church minister who authored a history of the Ojibwa community he served, described Netley Marsh as "ideally suited" to the community's needs: "Between Netley Creek and Lake Winnipeg the river created its own all-encompassing world. The delta area was a labyrinth of channels, lakes, islands and muskegs – a 'wilderness' to white settlers, but a homeland for hunting, fishing, and gathering people who traveled by water."[75] This description suggests how Aboriginal provisioning strategies were adapted to substantial wetlands such as Netley Marsh. In contrast to newcomers, who focused on small wet areas as agricultural problems to be solved, nineteenth-century Ojibwa were oriented to large wetlands they viewed as resources to be valued.

Through the creation of the province in 1870, Aboriginal people retained title to the region's lands, with the possible exception of some lands along the rivers and a small number of privately negotiated arrangements affecting small parcels of land.[76] For the dominion government, the Aboriginal claim was a pressing concern due to fears it would retard large-scale migration from Ontario, the United States, and overseas.[77] Only if the government had firm title to the lands of the northwest could settlement be pursued vigorously. For their part, Aboriginal people recognized the importance of securing their rights in advance of an influx of immigrants, and they were proactive in their efforts to negotiate with the government.[78] The first treaty between the Canadian government and the Aboriginal people of the prairie region was signed in August 1871.[79] Contested as it was, the document provided the basis for Aboriginal dispossession throughout much of what is now south-central Manitoba.

A key aspect of the treaty was Aboriginal reserves. In the federal government's conception, they were to be parcels of land set aside exclusively for Aboriginal use, though the land would remain under federal jurisdiction. The reserve system seemed to offer something like security of tenure, which Aboriginal people recognized they lacked across much of what had been their domain. But negotiations were difficult. Reserve size became a key point of contention as Aboriginal people sought a far larger amount of land than the government was willing to give them.[80] The dispute was resolved in part through a government commitment to tolerate hunting, fishing, gathering, and haying on land not immediately required for settlement.[81] Within a few years, however, increasing settler demand for land as well as more concerted government controls over resource procurement began to impinge on Aboriginal access to off-reserve resources.

Diminished access to off-reserve resources made it more important to increase returns from reserve lands. In the wet prairie, drainage was a key means of doing so. Soon after treaty, Aboriginal people began taking their plans for drainage to Indian agents, who conveyed them to the government. Typically, officials expressed no objections to the idea of drainage but were concerned about funding it. It was made clear that bands would be obliged to pay all the associated costs, including surveys, materials, labour, and engineering services. If the band lacked the necessary funds, then the project would not go ahead.[82] This was in contrast to how things worked between the provincial government and non-Aboriginal settlers, for Manitoba provided abundant supports to newcomers interested in drainage. Nothing like the generous grant from the province to the Rural Municipality of Rhineland would be forthcoming from the federal government for Aboriginal people.

Aboriginal difficulties in gaining assistance with drainage were partly a product of jurisdictional disputes between Ottawa and Winnipeg. Under the British North America Act, drainage was a provincial responsibility, whereas Aboriginal matters were a dominion concern. Continuing controversy over dominion ownership of Manitoba lands, an arrangement that ran counter to the precedent set in 1867 when the original provinces of Confederation retained authority over their own lands, established a context in which both governments kept a close eye on jurisdictional matters. By the late 1870s, the provincial government had undertaken many drainage projects and offered both grants and loans to settlers and municipalities interested in drainage. For its part, the dominion government was careful not to become involved in anything like a drainage project. In the words of one dominion official many years later, "we have never ... felt that we could step into a problem [drainage] which is the responsibility of the Province."[83] Aboriginal people who sought help with drainage were asking for something that the provincial government made available to their non-Aboriginal neighbours but that the dominion government was unlikely to grant for reasons beyond the validity of any particular request or even the racial identity of the petitioners. In their relationship with the federal government, Aboriginal people were enmeshed in a bureaucracy not tasked with responsibility for drainage.

Some Aboriginal communities did manage to begin construction of drainage works in spite of the significant financial barrier. Aboriginal labour was employed, and the results impressed government inspectors. In 1886, for example, Indian Agent McColl claimed that, "in all my travels throughout the province of Manitoba, I did not see anywhere such a perfect system of drainage as that performed by the Indians of the [St. Peter's] Band."[84] Unfortunately, those who had undertaken the work were unlikely to ever see the full benefits of their labours. As McColl explained, the resources of the reserve had proven insufficient, and work had been "abandoned before completion for want of funds."[85] Even with such a promising start, no government assistance was forthcoming. Partially constructed drainage works not only failed to provide the anticipated benefits but also could alter hydrological circumstances so as to exacerbate flooding. On the St. Peter's reserve, the dominion government not only refused to help continue the improvements but also was willing to leave the Aboriginal community vulnerable to risks derived from incomplete construction.

Despite unresolved problems with flooding, the lands of the St. Peter's reserve were especially desirable to newcomers because of their location immediately north of Winnipeg. In the early years of the twentieth century, as a prosperous period dawned on the prairie west and settler demand for land increased, Liberal Prime Minister Wilfrid Laurier spearheaded federal government legislation that facilitated the surrender of Aboriginal reserves.[86] Amid much controversy, the Aboriginal residents of St. Peter's were relocated to a new reserve on the eastern shore of Lake Winnipeg in 1907.[87] In their new location, inadequate drainage was also a problem. The new reserve included "much flat land that is not sufficiently drained," and the chief and councillors appealed for government support with ditch construction. Despite the initiative the community had shown, and aside from concerns about how to pay for construction, government agents doubted the value of undertaking such work. Although drainage "would no doubt enhance the value of the land and make it possible for more extensive farming operations," it was thought that there was "no guarantee that the improvement of the lands would be appreciated or made use of."[88] Here drain construction was forestalled principally by doubts about the capacity of those whom it would serve. Historian Sarah Carter has documented how government officials denied agricultural implements to progressive Aboriginal farmers.[89] In a manner consistent with her interpretation, drainage was one more tool that dominion officials were reluctant to make available to Aboriginal people.[90] Although federal-provincial jurisdictional arrangements certainly created barriers to Aboriginal drainage, dominion unwillingness to offer assistance was consistent with a pattern of stinginess justified by racist presumptions about Aboriginal agriculturalists. The result was that large-scale drainage was rarely undertaken on Aboriginal reserves.

The contrast between settler lands where drainage supports were available and Aboriginal lands where they were not was heightened as Manitoba drainage projects could have direct and negative consequences for reserves. A major concern in the sort of systematic drainage undertaken in drainage districts was with identifying a safe outlet for excess water, which would ensure that improvements to one area would not lead to flooding in another. Generally, this meant carrying water through to a stream, river, or lake that could absorb the flow. The Long Plain Indian Reserve, situated west of Winnipeg and south of Lake Manitoba, was located between the Assiniboine River and a 1906 drainage project undertaken in the Rural Municipalities of Portage la Prairie and North Norfolk by the provincial government.[91] The reserve community had not granted permission for a project that would certainly cut up their land and perhaps result in worsened flooding. The federal government protested the matter, arguing that it was unjust for the province "to drain the water of parts of two Municipalities into the Reserve, causing the flooding of all the good hay lands on the Reserve."[92] In this case, it was only federal action that kept an Aboriginal reserve from operating as an adjunct to the province's drainage geography, an area in which drainers could disregard the human consequences of changes to surface water patterns. Whereas the federal government was unwilling to offer support for Aboriginal drainage, the provincial government was disinclined to plan drainage projects in a way that respected the integrity of Aboriginal lands.

The ecological commons of agricultural flooding and the drainage designed to address it created common cause among many of Manitoba's cultural groups as well as a parallel orientation toward the expanding provincial state. The experience of Aboriginal groups was very different. They tried to leverage the state in a manner comparable to that of neighbouring non-Aboriginal groups but were largely unsuccessful. Ultimately, Aboriginal and non-Aboriginal lands were distinguished not only by the boundaries established at the moment of reserve creation but also by how land change in these areas occurred under substantially different political conditions. Jurisdictional divisions between the provincial and federal governments amounted to major barriers in the way of drainage on Aboriginal reserves. The situation was compounded by government officials' belief that Aboriginal people would not pursue progressive farming, with the result that reserve land would not be used as productively as non-reserve land. Such assumptions prompted governments to limit assistance for projects such as drainage and served to inscribe notions of race onto the landscape. The creation of reserves should perhaps be viewed as a long-term process bound up in environmental change related to power relations among human groups.[93] The process of transforming Manitoba's wet prairie changed much of the landscape of the provincial south but left unchanged or even harmed Aboriginal lands.

## CONCLUSION

Agricultural settlement under the provisions established by the federal government made more difficult Manitoba's task of creating local infrastructure such as road systems. Soon after settlement began in earnest, it became apparent that farmers in many areas were vulnerable to flooding. The province began to undertake agricultural drainage, but dissatisfaction with provincial initiatives led some settlers to embrace municipal government as a means to take control over locally necessary infrastructure projects. However, municipal drainage shared many of the financial and administrative problems of provincial drainage. A system of drainage districts was developed to provide a means of charging the costs of drainage to those who would benefit and to allow for a coordinated drainage effort across municipal boundaries. To understand the significance of drainage in Manitoba, it is important to consider the administrative structures and communities of interest that emerged from the tension between the settlement system enacted by the dominion government and the province's wet prairie landscape. Both administrative mechanisms and shared experiences spanned key cultural divisions, contributing to a provincial context that was formed in part in relation to challenging environmental conditions. The ecological commons of surface water and the drainage infrastructure created to address it had profound consequences for those who lived in the wet prairie.

Defined in relation to perceived flood patterns, drainage districts extended across the cultural divisions that historians have seen as fundamental to life on the Canadian Prairies. Although the federal government had allowed, through amendments to the Dominion Lands Act, distinctive patterns of landholding on Mennonite reserves, the environmental conditions of southern Manitoba and the provincial mechanisms established to address them served to create significant connections among neighbouring cultural groups. Flows of surface water linked different communities and served to reorient them, even those as traditionally insular as the Mennonites, toward the provincial government. Structures of authority differed substantially on Aboriginal reserves. Due to perceived barriers of jurisdiction and race, Aboriginal reserves remained largely insulated from the environmental and cultural outcomes of drainage. Studying drainage in southern Manitoba exposes the historically and geographically specific ways in which environment and culture intertwined. They were as intricately interwoven as the different landscapes in the "marsh, bog, ridge, marsh, bog, ridge" pattern that Henry Youle Hind encountered on his exploratory journey through the future province.

# 2

# Jurisdictional Quagmires: Dominion Authority and Prairie Wetlands, 1870-1930

Until a few decades ago, North American wetlands were notorious landscapes, at least among newcomers. Where malaria was a problem, swamps were thought to be the origin of the unhealthy vapours blamed for the illness. Manitoba's environment did not support the particular mosquito that was eventually confirmed as the transmitter of malaria. Yet even if there was only limited concern over the health effects of rural wetlands in Manitoba, there was widespread agreement that they were not useful areas. They amounted to blights on the productive agricultural landscape the township survey was intended to frame. Indeed, until the 1970s, when the functions and values of wetland ecosystems began receiving widespread attention, those committed to agricultural intensification generally dismissed wetlands as places of little utility.[1]

Yet beginning in the late nineteenth century and continuing into the twentieth century, Manitoba and the dominion fought over what they termed the waste lands of the province. To explain why both governments desired control over what their very language confirmed were not particularly useful areas, it is necessary to wade from the wet prairie into the equally murky terrain of federal-provincial relations. This chapter considers how the dynamic nature of the wet prairie landscape, evident in seasonal and annual variations in wetland size and shape, affected relations between the federal and provincial governments. Under the Manitoba Act, the 1870 government of Canada legislation that created the Province of Manitoba, authority over the lands of the new province, both wet and dry, was retained by the federal government. This was contrary to the desires of many who in 1870 lived in and around the Red River settlement (at the forks of the Red and Assiniboine Rivers), notably the Métis, as well as to what local leaders thought best for the region.[2] Historians have interpreted these events in several ways. Chester Martin took the view that Manitoba had been unfairly treated and even participated in the province's effort to secure a more generous settlement.[3] Gerald Friesen has suggested that prairie residents took pride in the national purposes to which regional lands had been put.[4] Recently, Jim Mochoruk has emphasized the colonial nature of the relationship between Manitoba and the dominion government.[5] Environmental factors were not much considered by these scholars, for whom natural resources administration was principally an economic and political issue.

The first two parts of this chapter examine how the dynamic nature of the wet prairie landscape bore on jurisdictional disputes. Many of the challenges faced by dominion and provincial administrators in their attempts to resolve the matter derived from two characteristics of wetland landscapes. First, wetlands varied through time, changing in size and shape in response to the amount of water in the surrounding ecosystem. Second, wetlands were hard to demarcate at any particular moment because they were not often bordered by a significant environmental discontinuity such as a riverbank or lakeshore.[6] Typically, the changes in soil and vegetation that mark wetlands are gradual, and materials and energy flow readily across the environmental gradients.[7] Not realizing the significance of these wetland characteristics, government administrators preoccupied with jurisdictional conflict arranged land transfer programs that were rendered more complicated by the difficulties of explaining change in wetlands or even of identifying definitively which areas were wetlands. Scholars have emphasized the importance of maps and numbers in efforts to control territory, underlining how they comprised a key means through which the state sought to organize and administer the landscape.[8] The situation in Manitoba illustrates the difficulties that ensued from the management of a landscape that, because of its inherent variability, was hard to represent in a definitive manner. The wet prairie environment helped to shape jurisdictional conflict, as officials with both governments scrambled to assert their authority over a dynamic landscape. Ultimately, the disputes between Ottawa and Winnipeg over Manitoba's waste lands illustrate how political and environmental complexity compounded each other.

Although eastern areas of the Canadian prairies received more precipitation than western areas, wetlands were found throughout the region, and variability was characteristic of all prairie wetlands. Drainage became a concern in Alberta and Saskatchewan by the early twentieth century. Even in this dry tabletop region, drainage could serve as a complement to irrigation and as a means of increasing lands available for settlement. As in Manitoba, there were serious jurisdictional impediments to drainage. An examination of the comparatively straightforward effort to circumvent these impediments not only offers a useful counterpoint to the Manitoba experience but also provides evidence of growing appreciation among dominion government officials for the importance of administering wet areas with an eye to their dynamic character. Not only mapping and counting but also the land management facilitated by these techniques would need to proceed differently in such landscapes. It was only after jurisdiction over Manitoba's waste lands was sorted out that dominion officials sought to establish environmentally appropriate management policy.

## Contexts and Precedents

Dominion ownership of Manitoba's crown lands was a persistent matter of concern for early provincial leaders. Manitoba politicians argued that, without a land base, they lacked a major source of revenue. Indeed, the activities of the provincial government were constrained by financial limitations. This constraint was evident in relation to public works, as much-needed drainage projects were curtailed, and in the very work of governing, as the province switched from a bicameral to a unicameral political system in part to trim costs.[9] Had crown lands been under provincial jurisdiction, Manitoba politicians would have had the option of managing them to generate revenue. Provincial politicians recognized this and blamed their financial woes on the dominion government. They lobbied persistently for the transfer of crown lands to the province.

Manitoba's position was not without precedent. Decades earlier, some American states had been in a similar position. Early in the nineteenth century, extensive wetlands in the Mississippi valley remained under federal authority. States lacked a financial incentive to undertake drainage, and it was unlikely that settlers would seek to acquire lands they would be unable to farm. Large-scale flooding in the 1840s generated greater interest in land drainage and flood protection. Representatives from Louisiana and Missouri led efforts to put the matter before the US Congress. They proposed that ownership of wetlands be transferred to the states. These lands could then be sold and the proceeds used to fund the drainage works necessary to make them suitable for agriculture.[10] Similar provisions were later applied to California and several midwestern states.

The American Swamp Land Act of 1849 (as well as later versions passed in 1850 and 1860) encouraged drainage but over time became associated with corruption and abuse. This forestalled drainage in most areas.[11] For instance, Indiana spent the proceeds of swampland sales on projects not directly related to land reclamation and was unable to complete projected drainage works. For its part, Illinois sold large parcels of wetland to speculators, many of whom failed to provide adequate drainage. The Swamp Land Act was a creative attempt to promote land drainage, but its many failures reflected the substantial obstacles to successful implementation of such a policy.[12]

Despite the manifold shortcomings of the American legislation, Canadian officials took inspiration from it.[13] In April 1880, the dominion government undertook to transfer federal lands to Manitoba in return for drainage through what became known as the drainage lands arrangement. The province would be granted title to the available even-numbered sections (thereby excluding lands reserved for the Hudson's Bay Company and the school endowment) in areas that it improved.[14] It was largely to take advantage of the drainage lands arrangement that the provincial government passed the 1880 Drainage Act, which targeted some of Manitoba's substantial wetlands. The arrangement provided a means through which the province could be compensated for the cost of undertaking expensive drainage projects on lands not at the time under provincial control.

The province's willingness to take on the project of drainage was not sufficient to convince the dominion government the work would be done well. Keen to have an independent report to confirm the descriptions provided by Manitoba government officials, in June 1880 the federal government assigned Dominion Land Surveyor Lindsay Russell to examine the current surface water situation and to comment on the province's drainage plans. In his report to Ottawa, Russell emphasized the preliminary nature of the planned work and the challenge of draining areas so "thickly covered by the heavy growth of prairie grasses, and reeds or rushes." He suspected that the results of drainage might be disappointing and cautioned that transfer of title should not be hastily made: "It would be well that the cession to the Province ... should follow the fairly demonstrated success of the drainage."[15] Dominion officials adopted Russell's skeptical view. When provincial officials sought transfer of the relevant

lands some two years later, believing they had fully satisfied their end of
the bargain, they encountered resistance from Ottawa.

The province had asked for transfer of nearly half the 391,280 acres of
land served by the 188 miles of drains that had been constructed.[16] Surveyor
J.W. Harris was contracted to undertake the inspections on which the
matter was to hinge. Harris had originally come to Manitoba as a dominion
land surveyor. Since his arrival, he had undertaken some projects on behalf
of the provincial government and various private concerns as well as the
dominion government. His established relationship with both governments
did not forestall the complications that arose when he submitted reports
of his inspections in 1882 and 1883. Harris made it clear that the marshes
"cannot of course be said to be thoroughly and completely drained by the
expenditure already made," but he allowed that "a very decided improvement ... on the condition of the land" had been effected. The work to
date had been sufficient "to induce the taking up of unappropriated sections for homestead in portions of the marsh which were formerly considered of no value."[17] According to Harris, the province had done enough
to merit transfer, even if surface water problems had not been entirely
eliminated in all areas.

His reports argued for the transfer but did so in a moderate tone. There
was nothing out of keeping with Russell's 1880 predictions: no talk of a
dramatic change from wet to dry, no overnight transformation from an
agriculturally useless to a magnificently productive landscape. Nevertheless,
the reports from Harris were dismissed by dominion officials as too lauda-
tory to be believed as well as too general to be useful.[18] Ultimately, the
trouble in completing the transfer derived more from the difficulty of
specifying precisely which area had been drained than from the question
of whether the drainage was sufficient. Since much of the land in question
lay in townships that had not yet been subdivided, it was difficult to provide
precise legal descriptions of the drained land.[19] Even in surveyed areas,
wetlands – places inconvenient for surveyors to measure and evaluate –
were often left outside the grid. Dominion land surveyors simply classified
several larger marshy areas as lakes, making it impossible to separate even-
numbered sections from odd. Both Big Grass Marsh (to the west of Lake
Manitoba) and St. Andrew's Marsh (to the northwest of Winnipeg) were
listed as examples of areas traversed as lakes. As the order in council au-
thorized only the transfer of the even sections, the imprecision interfered
with the transfer.[20] Notwithstanding its eagerness to secure land, the
provincial government found it difficult to provide the sort of report that
would satisfy the dominion authorities.

Given the relatively small amount of land in question, why was the dominion so exacting? The intensity of the dispute makes sense when it is understood that land acreage and quality were of secondary importance. Land mattered more because of the larger political context in which it was embedded than as an object in its own right. By dealing in wetlands, the governments addressed one of the environmental constraints on farming in Manitoba. But the focus on wetlands also provided a non-political means of transferring some land to the province. Both parties wanted to avoid setting precedents. The dominion did not want to transfer all of Manitoba's lands, and the province was loath to signal acceptance of dominion land ownership. By invoking an environmental logic, the dominion could make a transfer, and the province could accept it – without either side compromising its principled position. The dominion's insistence that the province meet the terms of the drainage lands arrangement was intended to maintain the jurisdictional status quo by which crown lands remained dominion property.

Yet because dominion demands were extremely difficult to fulfill, no mutually satisfactory solution was found. The situation degenerated, and both dominion suspicion and provincial exasperation increased as correspondence flew back and forth late in 1882 and throughout 1883. Eventually, the province marked on a map the acres that it had rendered fit for sale, Harris certified the map, and in March 1884 the dominion accepted the claim.[21] The province received about 62,810 acres (25,418 hectares) near Winnipeg and about 48,310 acres (19,550 hectares) in areas southwest of Lake Manitoba.[22] This anticlimax prompts a question: Why were both governments suddenly willing to find a solution? While in the course of time some government officials had argued over the transfer of lands under the drainage lands arrangement, others had fought over the rest of the lands of the province. By 1884, both governments were anticipating further negotiations concerning Manitoba's place in Confederation. Given the potential scope of these discussions, with the province asserting its right to the entire 94,888,467 acres (38,400,000 hectares) within the provincial boundaries of the day, the approximately 112,000 acres (45,324 hectares) affected by the drainage arrangement were put in perspective.[23]

Manitoba's wetlands had become a political instrument. As the political context changed, so did the importance of the drainage lands. It was not that the province had suddenly figured out how to record this challenging landscape but that the governments had both decided, in light of other pressures, to overlook the persistent problems with mapping the wet prairie. As it became apparent that the drainage lands arrangement had proven an insufficient panacea for provincial discontent, officials concluded that it was more important to move on than to continue to wrangle over an arrangement that (to Manitoba's satisfaction) had put some lands at the province's disposal but that (to the dominion's frustration) had failed to achieve its political purpose.

## THE SWAMPLANDS ARRANGEMENT, 1885-1912

Negotiations between Winnipeg and Ottawa culminated in an 1885 agreement that redefined Manitoba's position within Confederation. Premier John Norquay seemed to be well pleased with what became known as the Better Terms Agreement, which included an increase in financial support from the dominion. The agreement also provided for the transfer to the province of all crown lands in need of reclamation, estimated at between 7 and 10 million acres.[24] Manitoba would be able to generate revenue by draining and selling or leasing these lands. This provision was elaborated and formalized by the passage of the Swamp Lands Act later in 1885. Norquay saw the swamplands provision as a "main feature" of the arrangement and optimistically predicted that the 1885 agreement might "finally dispose of the land grievance question."[25]

Under the earlier drainage lands arrangement, lands were to be transferred after they had been drained. Now, under the Swamp Lands Act, lands were to be transferred as soon as it was established that they required drainage. The work of inspection and the transfer that hinged on it, which the controversy surrounding the reports from Harris on the drainage lands made clear could be problematic, would come before rather than after drainage. Although the drainage lands arrangement operated in relation to nine specified projects, the swamplands arrangement pertained to all wetlands in the newly enlarged Manitoba. The entire province would have to be inspected. By the late 1880s, swampland commissioners were appointed and assigned the enormous task of identifying all the swamplands in Manitoba.[26]

Quickly, it became apparent that this was a complicated and time-consuming endeavour. The province was large, and wet regions were far flung. Even the most diligent commissioners could travel only so fast by horse and carriage over rough terrain. The regulations governing their work, drawn up by politicians and bureaucrats, did not always suit conditions in the field. For instance, commissioners had to petition Ottawa for approval to inspect particularly wet areas while the ground was frozen and thus more easily traversed.[27] Interactions with local residents were protracted. And commissioners had to appeal again and again for adequate funds and basic supplies while Ottawa and Winnipeg bickered over which government should pay the bills.[28] The nature of the task and the many logistical problems meant that it took a long time to determine which lands should be transferred to the province under the Better Terms Agreement. Indeed, inspections continued well into the twentieth century. Premier Norquay had not foreseen this when he expressed satisfaction with the 1885 agreement. In effect, Manitoba was left waiting for its better terms.

Deliberate delays by dominion officials were partly to blame. Since the land remained the property of the dominion until it was formally transferred by an order in council, the government in Ottawa continued to derive revenue from timber and grazing leases.[29] Moreover, the dominion was trying to dispose surreptitiously of whatever swamplands it could while still retaining the right to the revenues. This is made clear through two letters from H.H. Smith, commissioner of dominion lands, to A.M. Burgess, deputy minister of the interior. Both are dated 6 August 1889. In the two letters, Smith argued that the dominion should sell the land it then administered under hay lease because of the trouble and expense of managing the leasing system. One letter then continued that "there is a reason for selling these lands now which I cannot very well advance officially. It is altogether likely a good many of them may, before long, be selected by the Swamp Lands Commissioners and we shall lose them entirely. We had better, therefore, make what we can out of them while we have the chance."[30] Whether or not they had any knowledge of the devious strategy recommended by Smith, provincial officials were incensed by dominion delays. Not only was Manitoba left waiting for its better terms, but also as the years passed the better terms became less good.

Provincial officials recognized that not all the blame could be laid at the dominion's door. Ongoing environmental change not attributable to human activity was complicating matters. Although the drainage lands arrangement indicated the difficulty of establishing boundaries in a wetland landscape at a particular moment in time, the swamplands arrangement made clear that the temporally dynamic nature of the wetland landscape was equally problematic. As wetland ecologists G. Mulamoottil, B.G. Warner, and E.A. McBean have explained, "even small changes in surface and ground water hydrology may result in significant changes to the wetland."[31] With land ownership hinging on land condition, ongoing environmental processes were invested with political significance.

For some years after 1880, the region trended toward dryness, even as intermittent wet years increased demand for drainage. Manitoba officials recognized that the lands of the province were "becoming dry and changing in character."[32] Although drier land improved agricultural prospects in some areas, an outcome that pleased all with an interest in the development of the Canadian west, the provincial government saw a downside. Land that dried naturally prior to inspection by the swamplands commissioners remained the property of the dominion; land that had to be drained became provincial property. Thus, natural environmental improvement meant a territorial loss to the Manitoba government. Insofar as the province hoped to sell or lease these lands for profit, the loss was also financial.

Manitoba officials also found themselves in an awkward position in relation to artificial drainage. Since drainage was a provincial responsibility, politicians had to respond to settlers' demands or face voter dissatisfaction. Particularly in light of difficulties attracting and retaining settlers in Manitoba during the 1880s and early 1890s, ditch construction could not wait until the swamplands commissioners finally completed their inspections. But artificial drainage, supported by provincial money and policy, decreased the amount of land that qualified as swamp and that was therefore eligible for transfer to Manitoba.[33] The province tried to make the best of the situation by taking steps such as assigning an inspector to accompany the swamplands commissioners. The inspector was to draw the commissioners' attention to any land that had been swamp in 1885 but that had since been drained through local efforts.[34] Nevertheless, provincial officials remained uneasy, fearing both dominion cunning and environmental change.

The annual reports of the Department of Provincial Lands, which had been established in 1888 in part to manage the lands the province anticipated receiving in transfer, revealed Manitoba's mounting frustration. Hoping to spur more rapid transfer, the province used the question of natural resources ownership as leverage. In 1890, Commissioner Joseph Martin railed that

> it has always been contended by the Dominion Government that one great reason that they insist upon administering the lands is that they are able to do it so much better than a local government. This Department has no hesitation in claiming that work of this kind could be done in one-fiftieth of the time that it takes your Department and we have only a staff of one man.[35]

Designed to mitigate intergovernmental disputes over land ownership, the Swamp Lands Act compounded tension as it became a source of dispute itself. An arrangement designed to appease Manitoba became ammunition for provincial rights advocates. In the view of Manitoba's provincial lands commissioner at least, "great loss and injury [are] being sustained by the province by the failure of the Dominion government to carry out the arrangement entered into in 1885."[36]

Dominion officials acknowledged that changes in wetland size and shape were problematic. Soon after the 1885 agreement, they decided that lands subject to flooding in an average year (rather than lands inundated in the year of inspection) would be transferred. From then on, swamplands commissioners were expected to make allowances for annual variations.[37] When a perceived trend toward a drier environment caused Manitoba to protest, the dominion agreed that the amount of land to be transferred should equal the amount that had been wetland in 1885.[38] Both policy adjustments were attempts to cope with changes over time. Yet incorporating new instructions and compensating for environmental dynamism further slowed the inspection work of the commissioners, and, as the inspection process took longer, the likelihood of significant change in environmental conditions increased.

The swamplands commissioners recognized the awkwardness of the swamplands arrangement and communicated their concerns to the dominion government. Commissioners William Wagner and William Crawford warned the dominion to guard against settlers who saw an opportunity to gain by denying the existence of swamps. The Canadian government made land available to newcomers at remarkably favourable terms, whereas the province sought to manage its land for profit.[39] After swamps had been transferred and drained, Wagner and Crawford warned, it was possible that settlers might claim that the land had always been dry and thus should have been dominion land available as free homesteads rather than provincial land offered for sale.[40] The worry was that, if such a situation were to arise, the only way to guard against both public outcry and provincial discontent would be for the dominion to allow homesteading in these areas and to compensate the province for the loss of lands already identified as suitable for transfer. Clearly, this was not a desirable outcome for a national government keen to minimize expenditures. Wagner and Crawford also saw an opportunity for the province to profit through fraud. They warned that it was "in the power of the Government of Manitoba to create swamp lands at any point in the province where there are lands still at the disposal of the government of Canada which they wished to have handed over to the province."[41] In this view, flooding land could become a means of claiming territory, regardless of the consequences for settlement and agriculture.

Political control over lands and the environmental condition of the lands were themselves complicated issues that became only more challenging by how they bore on each other.

Both the federal government and the provincial government felt vulnerable in light of environmental changes and the possibility that the other would gain the upper hand through more successful adaptation. The environment was not the politically neutral arbiter that politicians had taken it to be. The situation was conducive to the outbreak of small but intense political battles. The resulting anger and frustration undermined both the larger political purposes (the improvement of intergovernmental relations) and the original environmental aims (the production of land suitable for agriculture) of the Swamp Lands Act. The perversity of the situation was epitomized in an 1893 exchange between Manitoba and Canada. The provincial government had produced a pamphlet extolling the virtues of Manitoba's lands and advertising those available for purchase. It read, in part, "what are known as 'swamp lands' are being conveyed by the Dominion Government to the Local [Provincial] Government. Many of these lands are not swamp lands at all, but are valuable for farming purposes."[42] A dominion official responded indignantly:

> You will observe that the paragraph contains the very important statement that *many of these lands are not swamp land at all*. Such being the case, I would be glad to have a list of these lands, and in view of the fact that the Manitoba Government are only entitled to the lands described as "swamp lands," if any others were by oversight conveyed to the Province, no doubt it is only necessary to call our attention to the fact in order to have them re-conveyed to the Crown.[43]

Increased immigration and settlement were desired by both governments. But population growth imposed heavy responsibilities for infrastructure construction and service provision on the province, even as it led to increased importation of goods, which inflated the dominion's tax revenues.[44] The dominion had the most to gain and the least to lose from immigration but still quibbled with the claims of provincial settlement boosters. Typically, officials got hung up on small details and lost sight of the larger goals of the swamplands arrangement.

The arrangement was an aspect of the 1885 Better Terms Agreement, intended in part to solve the complicated political problem of crown lands in Manitoba. While disagreeing over which government had the better claim to provincial resources, both concurred that Manitoba's lands could be divided into those fit for settlement and those in need of drainage. The basic division between potentially settled areas and frequently sodden lands acquired additional meaning as it was incorporated into intergovernmental negotiations on land ownership. However, governments invested wetlands with political significance without adequately anticipating or effectively accommodating the environmental processes at work in these areas. Locating these lands on the ground and then recording them on a map proved to be intensely problematic undertakings. As a result, the use of Manitoba's wetlands as a political vehicle drove the province and the dominion further into a jurisdictional quagmire.

As the marshes of Manitoba changed and were changed, trouble stretched on into the first decade of the twentieth century. Through decades of intergovernmental confusion, it became apparent that it would be impossible to bring the swamplands matter to anything approximating a logical conclusion. In this context, an abrupt policy shift became increasingly probable, and early twentieth-century political developments made it even more likely. The 1911 federal election brought the Conservative government of Robert Borden to power. The Borden campaign had been greatly assisted by Conservative Manitoba Premier R.P. Roblin, and this cooperation set the stage for a new round of negotiations over Manitoba's place in Confederation. In 1912, Manitoba accepted what historian Jim Mochoruk has described as a valuable package that included lump-sum payments of $202,723.57 to construct public buildings and a significant boundary extension. But the agreement also demanded a number of concessions from the province, including the return of the unsold swamplands to the dominion.[45] Although some swamplands had been sold at good prices, the majority were in the less fertile and more isolated northerly and easterly sections of the province.[46] Some 2,131,006 acres (862,388 hectares) had been transferred to the province; some 1,164,412 acres (471,220 hectares) were returned.[47] From one angle this seems to be a hasty and unexpected conclusion to a matter that had drawn on for decades; from another, it seems to be an appropriately political ending to a policy that had faltered largely because it had not adequately accommodated the dynamic nature of Manitoba's wetland landscape.

RECLAMATION ON THE PRAIRIES

Dominion interest in Manitoba drainage declined after the return of the swamplands in 1912. At about the same time, the federal government

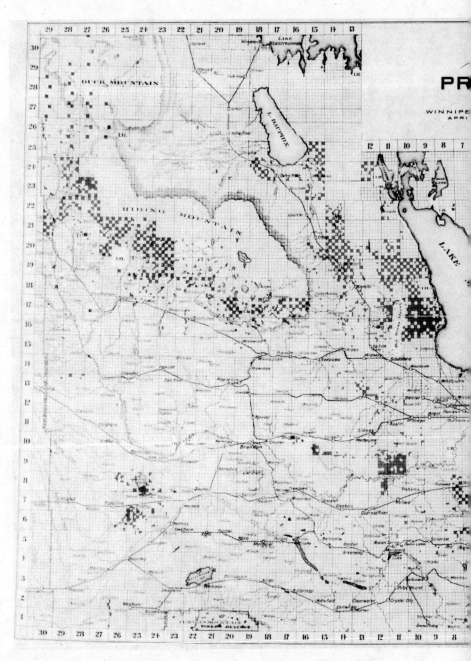

FIGURE 8   Map of Manitoba showing provincial government lands for sale, 1900. This map is one of a series prepared to facilitate the sale of lands transferred to the province as drainage lands or swamplands.

*Source:* Archives of Manitoba, Manitoba, map, Manitoba Provincial Government Lands for Sale, 1900, H7 614.2 gbbd series 1 1900 (CNeg 6802).

became more involved with surface water management in Alberta and Saskatchewan. The reasons for this involvement go back a decade or more before the 1905 creation of these provinces and turn in part on the contributions of a particularly energetic and influential civil servant. William Pearce was one of two officials appointed to the Dominion Lands Board, which had "responsibility for making regulations, recommending legislation, formulating resource development policies, and supervising the exploitation of all land, timber, minerals, and water resources throughout the Northwest."[48] As the inspector of land agencies, Pearce drew on his administrative talents and his surveying experience to organize and supervise land offices across the Prairies.[49] Described by one historian as the dominion troubleshooter, Pearce re-evaluated established practices and engaged in policy development.[50]

In his policy work, Pearce found inspiration as much among the work of his contemporaries as in the condition of the Canadian Prairies. The activities of key figures such as John Wesley Powell and Elwood Mead, both Americans prominent in water management circles domestically and internationally, caught his attention.[51] As noted by numerous environmental historians, Powell was an early promoter of innovative environmental management. He believed that water in the arid American west should be apportioned to ensure the efficient and equitable use of a scarce resource. Powell took the drainage basin as the key management unit, with water use to be determined in relation to basin geography. In his view, government management was necessary to ensure that administration served the public interest. Although neither unprecedented nor uncontroversial, Powell brought such thinking to the American government, and his contributions have influenced many, from managers and citizens to geographers and historians.[52]

By the early 1890s, William Pearce was convinced that the Canadian government should establish a legislative infrastructure appropriate to watershed management as promoted by Powell.[53] Despite government reluctance to take any action that acknowledged an actual or potential shortage of water on the Prairies for fear it would deter potential settlers, and particularly in the context of a series of dry years that even the government could not simply ignore, the weight of opinion gradually lined up behind Pearce's proposals.[54] The result was the passage in 1894 of the Northwest Irrigation Act. Under this statute, all water not already acquired by an act of Parliament (as in the case of small irrigation companies operating under a federal charter) or reserved by prior appropriation (as in the case of riparian rights claimed by early settlers) was declared to be the

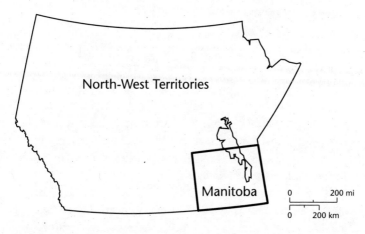

FIGURE 9    The northwestern interior of North America in 1894
*Source:* Adapted from Gerald Friesen, *The Canadian Prairies: A History* (Toronto: University of Toronto Press, 1984), 90-91.

property of the crown.[55] Under these terms, all water in the North-West Territories could be accessed only through a permit system administered by the dominion government. This legislation modified one of the basic principles of the British North America Act, under which the dominion government had jurisdiction over large navigable waterways, whereas the provinces administered smaller, non-navigable streams. According to historian C.S. Burchill, this act represented nothing less than a "radical" expansion of dominion authority.[56] The Northwest Irrigation Act certainly did change the legislative context for water management in the area it affected. In correspondence with William Pearce, Elwood Mead suggested that such legislation was progress, explaining how he had argued publicly that "Canada was in advance of the United States in its irrigation laws."[57]

The provisions of the Northwest Irrigation Act were not modified in 1905 when Alberta and Saskatchewan were created, so the dominion government had more control over waterways in Alberta, Saskatchewan, and northern Manitoba (which, as part of the North-West Territories in 1894, was subject to the legislation) than in the small, southerly rectangle that was then Manitoba (see Figure 9). After 1894, water management was different in the agricultural regions of Manitoba, which were largely situated in the provincial south, not only because of relative abundance but also because of legislative context. The Northwest Irrigation Act was designed to provide greater dominion authority over water in areas where irrigation was necessary, which suggests that a rough environmental logic underpinned the exclusion of the wet prairie. However, the act applied even to the wetter, more northerly reaches of the later province of Manitoba but not to the dry corner of the southwest, which is environmentally comparable to southeastern Saskatchewan. This suggests that the boundaries of the Northwest Irrigation Act might have been influenced by authorities' desire to avoid the enduring dispute surrounding Manitoba's provincial rights. Had the dominion government attempted to claim non-navigable waters in Manitoba, it would have been obliged to annul an authority granted nearly twenty-five years earlier to a province with which jurisdictional relations remained highly contentious.

According to historical geographer Matthew Evenden, irrigation in Canada was distinguished by the dominion government's cancellation of riparian rights prior to extensive settlement and by its efforts to attract private investment rather than undertake projects directly.[58] This approach proved to be relatively effective in spurring development, as railway and land companies invested substantial time, money, and effort in irrigation. However, despite determined efforts by some such as Pearce to improve on the American system, Canadian irrigators encountered problems of their own, ranging from difficulties with convincing farmers and the government that irrigation was worth the cost to issues with managing the muskrats and beavers that flourished in the drained landscape.[59] Particularly significant among the various challenges for irrigators was how a division of powers appropriate for irrigation became problematic when attention turned to drainage.

In the dry prairie, drainage was necessary largely to cope with the problem of seepage.[60] Irrigation directed water, but it did not ultimately control it, and much water that percolated away collected in lower areas. "In almost all large irrigation projects, particularly if they have been in operation for some time," explained a memorandum circulated in 1913 in the Water Resources Branch of the Department of the Interior, "the lower-lying lands, which when water was first delivered upon them were the most valuable lands within the tract, became after a time so water-logged as to be unfit for cultivation." The situation was even worse than the government memorandum made clear: elevating the local water table not only resulted in excess water but also raised salts from the subsoil, contributing to a salinization process that was incompatible with plant growth. To maintain agricultural production in irrigated areas, drainage was, as the author of the memorandum recognized, "as necessary in connection with the successful operation of an irrigation tract as is the delivery of water."[61]

Although immensely important as a means of coping with the environmental consequences of irrigation, drainage in Alberta and Saskatchewan was not merely an adjunct to irrigation. Eliminating wetlands, both the small prairie pothole variety and some more substantial wetland complexes, seemed to be an important means of supporting agricultural intensification during the first decades of the twentieth century, a period of extremely rapid settlement in the Canadian west. Drainage, such a major concern throughout Manitoba's wet prairie, was also an important issue in some areas of the dry prairie of Saskatchewan and Alberta.

It was an irony. Despite the extensive infrastructure necessary to provide sufficient water for agriculture in many areas of Alberta and Saskatchewan, some farmers and settlers in these provinces were troubled by excess surface water. It was also a frustration, as drainage advocates in the more westerly provinces faced legislative impediments that Manitobans did not. As with Manitoba in 1870, Alberta and Saskatchewan were granted jurisdiction over land drainage in 1905. In contrast with Manitoba, they lacked legislative authority to undertake or approve work that affected any stream, river, or lake. Since drained water had to be routed somewhere and would inevitably alter any river or stream to which it was directed, dominion control over all alterations to any water body under the 1894 Northwest Irrigation Act effectively forestalled drainage initiatives by the provincial governments. A division of powers appropriate to irrigation became troublesome once drainage was a concern.

The situation was very different in Manitoba, where the provincial government had authority to alter non-navigable waterways. It could allow, prohibit, or ignore changes to drainage patterns made by individual settlers, government agents, or private contractors as they sought to improve the lands of the province. Manitoba's control over non-navigable waterways was a rather circumscribed sphere of authority, but it was critical for drainage. The practical work of drainage in Manitoba was maintained on a relatively independent trajectory, one that was substantially different from that in the two more westerly prairie provinces, partly because the provincial government was responsible for the small waterways that most often figured in drainage works. The drainage and swamplands arrangements with the dominion government were important – if problematic – incentives and supports, but the province undertook a substantial amount of drainage entirely under its own auspices.

Although able to authorize alterations to non-navigable waterways in southern Manitoba because the area was not subject to the Northwest

Irrigation Act, Manitoba was as constrained as Alberta and Saskatchewan when drainage required alterations to navigable waterways, which in all three provinces were subject to dominion authority under the British North America Act. Some of the most severe chronic flooding in Manitoba occurred in the agricultural lands along the length of the Assiniboine River. Despite the extreme variability of its flow, it was considered navigable. As a result, efforts to address water problems along the Assiniboine were forestalled in some instances, even as drainage projects that did not directly affect the river went ahead in nearby areas.[62] Drainage initiatives along the Red River do not seem to have been forestalled by its status as a navigable waterway. This may be attributable to the fact that many early drainage ditches dumped into small natural creeks or streams that ran a short distance back from the river rather than directly into the river itself.[63] This was the case even with major drains and floodways. For instance, the Shannon Creek Spillway, a large drain running through Drainage District No. 2, "emptied into the Morris River by way of the Moyer (Lewis) Coulee."[64] Draining into tributaries rather than the main stem of the Red River might have circumvented jurisdictional conflict in some cases. But it is also possible that Saskatchewan and Alberta were unnecessarily cautious about trespassing on dominion jurisdiction through drainage work. Indeed, the dominion was intensely displeased with its responsibilities for flood mitigation along the Assiniboine River and might have been willing to tolerate provincial incursions to prevent the development of similar situations.[65]

With respect to drainage, Manitoba found ways to act more or less unilaterally; Alberta and Saskatchewan did not. As a report of the Reclamation Service of the Department of the Interior neatly summarized, in the two more westerly provinces "divided jurisdiction prevented any material development of drainage either by the Dominion or the Provincial Governments and invited controversy between them."[66] The dominion was willing to concede that its absolute authority over bodies of water under the Northwest Irrigation Act was not always conducive to progress. Small marshes and shallow lakes might "serve no useful purpose as sources of water supply" and could pose "a serious detriment to the districts in which they occur."[67] Officials from these provinces and the dominion realized they had a problem, but solutions were not obvious. Intergovernmental negotiations ensued.

Through a series of conferences, meetings, and correspondence that began in 1914, Alberta, Saskatchewan, and Ottawa divided responsibility for drainage between governments, based on project size and location. As outlined in the *Journal of the Engineering Institute of Canada* in March 1919, drainage was separated into four classes: small drainage projects, drainage in connection with road construction, drainage of dominion land in organized drainage districts, and drainage work initiated by the dominion government.[68] Each class had a distinct procedure through which funding and approval were to be obtained. Along with supporting legislation passed at both the provincial level and the federal level, these regulations provided an administrative infrastructure tailored to the jurisdictional arrangements governing surface water in Alberta and Saskatchewan.

Only in considering how the regulations were applied does it become possible to assess whether they were equally appropriate to environmental conditions. The federal regulations came into effect in the mid-1910s, just as a series of dry years began. By 1919, many water bodies in Alberta and Saskatchewan were much diminished or even completely dried up. The drainage that had seemed to be so pressing now seemed to be irrelevant in many areas. From figuring out how to facilitate artificial drainage, dominion officials switched to considering the management of lands that previously had been covered with water. Those tasked with examining the situation concluded that "the natural recession of lakes and sloughs in the west is influenced by two main factors." The first was "recession due to the fact that the cultivation of the land comprising the tributary drainage basin ... has reduced or cut off the natural run-off into the lake or slough." Since this change was seen as permanent, "the land so unwatered may be deemed to be dry land ... and may be dealt with in the same way as any other Dominion land."[69]

The second factor was "periodic recession dependent upon the natural precipitation and run-off." Since this change was not considered to be permanent, the land thereby exposed had to be managed with an eye to the possibility that it would again be flooded. Dominion officials recognized the dynamic nature of the wetland landscape. They were aware of the need to consider carefully changes in the environment and to accommodate the possibility of modifications in land management. With regard to much of the land newly exposed after the dry period of the late 1910s, the Department of the Interior thought that "it may safely be assumed that the natural condition of the land is such as to be unfit for cultivation," so "the Department would not be justified in dealing with it other than as a water-covered area not subject to disposition by patent until reclaimed."[70] Land that was only temporarily dry would soon return to its natural water-covered condition, and its administration should not be altered by intermittent dry periods.

Although dominion officials believed that this was sound administrative strategy, they were aware that it could leave them vulnerable to criticism from land seekers interested in the newly exposed land. They thought they were put in "a rather awkward position" by the failure of the public to differentiate between land that had become dry permanently and land that soon would be once more covered with water. They foresaw that settlers would "make every effort to secure the areas at present dry and will resent any action tending to interfere with their wishes in this respect." But at the same time, administrators were confident that, if they made the land available, it would be "only a question of a few years" until those who got the land would begin to protest "that the Department has sold them land which is either useless, or requires considerable expenditure for drainage to make it of use."[71] The risk was that the government might be put in a position where public outcry would oblige it to undertake difficult and expensive drainage projects.

Recognizing that only time would tell if the lands were permanently dry, officials in the Reclamation Branch recommended that newly exposed lands should be open for homesteading only if they had been dry for at least three years. If observers could not attest to this, then it was likely that the land would become wet again and consequently should not be available to prospective settlers. This was an attempt to adapt the state's administrative maps to reflect the possibility of change in land condition over time, an attempt to offer a temporal perspective on the question of whether or not a parcel of land was suitable for farming. Despite the frustrations of settlers who wanted the dried lands the dominion was withholding, at least some aspects of the proposed administrative arrangements seem to have been sensible. From a jurisdictional gridlock that had impeded drainage, Alberta, Saskatchewan, and Ottawa had developed a workable drainage policy. Even more significantly, officials exhibited an understanding that the variability of prairie wetlands should affect land management. Eagerness to secure newly available land would not be allowed to outweigh the caution necessary in a dynamic landscape.

Ultimately, drainage officials advised that the dominion should retain ownership of the highly variable wetlands in the southern parts of Saskatchewan and Alberta. This recommendation amounted to an effective and appropriate adaptation based on increasing understanding of a variable environment – precisely the sort of understanding that seems to have been absent among those involved some years earlier in administering Manitoba's swamplands. Drainage officials might have been motivated to protect settlers and guard against government liability, but they were also inspired by shifts in thinking about the value of wet areas. Such areas took on political implications in Manitoba before such thinking had significantly influenced government officials. By the time dominion officials addressed drainage problems in Alberta and Saskatchewan in the mid-1910s, administrators were aware that a marshy area could be valuable bird habitat.[72] Although it would be decades before the functions and values of wetlands were appreciated in full, it was already becoming more difficult to dismiss these areas as waste lands. Pressure from conservationists concerned with birds might have helped to keep drainage administrators focused on the environmental context for their work, making it less likely that political considerations would define policy.

As the variable nature of the Manitoba hydroclimate compounded jurisdictional disputes, so politically entrenched positions might well have forestalled the emergence of more sophisticated management of some wet areas in Manitoba. Shortly after Manitoba's swamplands were returned to the dominion in 1912, an employee of the Department of the Interior was asked to investigate the possibility of dividing these lands into two categories – "those which will require drainage and those which are fairly fit for settlement as they stand."[73] There were a number of legitimate means by which the province could have received good land under the swamplands arrangement. Square areas (quarter-sections or quarters of them, depending on the date) were classified as swamp according to the condition of the bulk of the land; 160 acres (65 hectares) legally defined as swampland and transferred to the province could, for example, contain up to 79 acres (32 hectares) of good land. Also, the dominion government had on occasion agreed to transfer good land to the province in lieu of swampland to meet the needs of settlers and railway companies that wanted to acquire land that had been classified as swamp. When the land was transferred back to the dominion in 1912, it was estimated that at least 10 percent of swampland was in fact not swampy at all.[74]

But how to distinguish this 10 percent from the rest? Through the work of the swamplands commissioners, Manitoba land inspectors, and a 1912 inspection undertaken by dominion agents, officials had a substantial body of information concerning the swamplands. But this information shared the limitations of the maps produced by dominion land surveyors. As one official explained,

> If the subdivision survey were made in a wet year, or after a series of wet years, or in the early part of the season, the surveyor probably found lakes, sloughs, or marshes, which he accordingly showed on his plans, which

became the Departmental record. If the survey were made in a dry year, or after a series of dry seasons, or late in almost any season, many of the smaller lakes or sloughs were dry, or practically so.[75]

In Manitoba, as in Alberta and Saskatchewan, lands that vacillated between wet and dry were a concern. If lands that were only temporarily dry were granted to settlers, then the inevitable return of water would bring a torrent of complaints. Caution in land disposal was the best way to limit future liability of the government. Responsible policy would take into account the inherently dynamic nature of the wetland landscape.

With the 1912 revestment of the unalienated Manitoba swamplands, the significance of the line between wet and dry changed dramatically. Transferring wetlands to Manitoba had provided a means of sidestepping jurisdictional barriers to drainage. But it had also allowed the dominion to appease provincial rights advocates without conceding that Manitoba had a legitimate claim to its own resources, for the transfer could be explained in environmental rather than political terms. The political consequences of land classification thwarted meaningful attempts to come to terms with actual environmental conditions in the province's wet prairie. Until 1912, any change from wet to dry or dry to wet produced aggravation and suspicion because of its implications for government jurisdiction. It was only after swampland management was separated from the continuing controversy over natural resource management that the challenges of administering Manitoba's dynamic environment could be addressed directly.

At numerous points during the late 1910s and early 1920s, Ottawa, Alberta, and Saskatchewan tried to involve Manitoba in their deliberations on drainage, with the aim of establishing a more consistent legislative regime across the region. Manitoba ignored most overtures, though its reasons were never made entirely explicit. Most likely, provincial officials simply found very little that spoke to their situation in these negotiations. Surface water management in Manitoba had followed a different trajectory. By the time Alberta, Saskatchewan, and Ottawa had worked out a legislative framework for drainage, Manitoba had launched a royal commission to investigate aspects of the substantial drainage system already servicing parts of the province. The Manitoba Drainage Commission was established by an order in council of 17 January 1919.[76] The investigations of this commission were extensive, reflecting the human and physical complexities inherent in Manitoba's drained landscape by this point. With regard to intergovernmental arrangements bearing on wetlands management before the 1930 transfer of natural resources to the prairie provinces, Manitoba was a very different place for both political and environmental reasons.

<h2 style="text-align:center">CONCLUSION</h2>

In his influential book *Seeing like a State,* James Scott compared a cadastral map to "a still photograph of a current in a river."[77] This is an apt comparison that conveys how state property maps fail to capture the dynamism often evident on the ground. In Manitoba's wet prairie, the role of water was more than metaphoric. The province's variable surface water geography, combined with arrangements under which control over lands hinged on land conditions, exposed the inability of the governments of Manitoba and Canada to administer effectively an environment that varied in character from season to season and year to year. With maps as their key tool, even well-meaning, hard-working government agents were not able to record anything beyond a moment in time. Eventually, a similar situation in wet areas of Saskatchewan and Alberta led to an arrangement under which land conditions would be monitored over a period of years before any decision would be made about how any potentially problematic parcel should be used. Although this was in large measure an attempt to protect the government from liability in cases of flooding, it amounted to a system better able to cope with environmental variability. In an effort to achieve more successful land management policy, government agents stretched their recording system to encompass change over time.

The challenges of wetland identification were certainly not unique to the Canadian Prairies or to the historical period here under consideration. Indeed, striking connections can be made between controversy over political boundary making in nineteenth-century Manitoba and the contemporary United States. In a collection of works of environmental philosophy, Edward Schiappa examines recent controversy in the United States over wetland delineation. He documents how overt conflict between those who would protect wetlands and those who favour development has been circumvented through political endorsement of a geographically limited conception of wetlands. Although scientists emphasize that wetlands expand and contract in relation to the amount of water in the surrounding ecosystem and explain that such processes are fundamental to ecological functions, political and corporate interests have favoured definitions limited to permanently inundated areas, because hiving off riparian zones makes more land available for development.[78] The parallel suggests how wetlands, literally situated along the margins of valuable dry land and useful water networks, have been, for a lengthy period and in diverse locations, at the centre of many critical debates over environmental management. As wetlands are transitional areas between land and water, so do they mark a particularly important intersection between political and environmental landscapes.[79]

Appreciating this intersection enhances understanding of a theme of long-standing concern to Canadian historians: jurisdiction over the lands and resources of the Canadian Prairies. Manitoba's drainage and swampland arrangements were meant to sidestep jurisdictional barriers to drainage and to ease contention over dominion ownership of provincial resources, but they proved to be a source of confusion and controversy. Administrators were preoccupied with the political consequences of land classification. Environmental change was evaluated primarily for how it bore on land ownership, and suspicion and hostility flowed from the dynamism inherent to regional wetlands. In this way, the drainage and swampland arrangements exacerbated the political conflicts they were partly intended to ameliorate. Drainage in Alberta and Saskatchewan was addressed at a later period and without the same implications for government authority. Dominion officials developed land management practices attuned to variability and the complications that might result if settlement went ahead in areas that were only temporarily dry. Only once Manitoba's swamplands were revested in the dominion government was there significant evidence of efforts to manage these areas in relation to their dynamic character.

Paying attention to wet areas sheds new light on the history of the prairie provinces by illuminating some of the key legislative and environmental discrepancies within the region. As Reclamation Branch head F.E. Drake wrote of the Northwest Irrigation Act, "the term 'Irrigation Act' does not adequately indicate the nature of the law." This was because

The Act vests in the Crown, in the right of Canada, the ownership and control of all sources of surface water supply within a certain described territory. It prescribes the purposes for which grants of the right to the use of water may be made, these being domestic, municipal, industrial, irrigation and "other." It is thus apparent that the Act covers much more than irrigation rights, and might more properly be described as a general water law.[80]

Environmental conditions and political developments operated in tandem to distinguish Manitoba from the more westerly prairie provinces. Although the exclusion of southern Manitoba from the Northwest

Irrigation Act was certainly not the only factor distinguishing the area, the need to disaggregate the prairie region is underlined through recognition that Manitoba's wet prairie was beyond the reach of what Drake called the dominion's "general water law." The pattern of state involvement in surface water management was significantly different in Manitoba. Historian Gerald Friesen has argued that the dominion's role in civil administration contributed to the consolidation of the prairie imagined community. "Ottawa treated the west," he has asserted, "as a single administrative unit for settlement, for lands and forests, for naturalization and police and Indians and transportation and the tariff."[81] But water policy was not applied so consistently. If the emphasis is shifted from land to water, then southern Manitoba might be better understood as a region unto itself.

# 3

# Drains and Geographical Communities: Experts, Highlanders, and Lowlanders Assess Drainage

On 22 February 1922, farmer L.C. Wilkin appeared before a committee of the Manitoba government. He explained how, in a district around the Rural Municipality of Morris that had been successfully farmed for over forty years, farmers were now experiencing very challenging conditions. According to Wilkin, surface water conditions had become worse over time, and many farmers were now faced with the difficult decision of whether to abandon their lands. Although farmers were bound to the area through hard work and social ties, the situation was so bleak that, as Wilkin explained, "ruin stares them in the face if this condition continues." Kirk, who had farmed in the area for thirty-two years, described how the floodwater now "comes much quicker and there is more of it." With regard to drainage, Kirk thought he and his neighbours were "paying for something we haven't got."[1] At minimum, Wilkin and Kirk were seeking financial assistance for stricken farmers. More broadly, they were expressing profound dissatisfaction with changed surface water conditions. Both thought the drainage district in which their farms were situated was not adequately protecting their lands.

The 1895 Land Drainage Act specified that drainage districts were to be designed by the professionally trained engineers and surveyors of the Department of Public Works, which would presumably result in projects that, in comparison with earlier undertakings, were more conceptually and structurally sound. From the progressive era onward, engineering advice increasingly came to carry the promise of greater efficiency through centralized planning and expert knowledge. But with regard to Manitoba drainage, as with other public works projects elsewhere, patronage continued to bear on which projects were undertaken and on how the work went ahead. By the mid-1910s, even though drainage district design and other factors such as climate variability and land use changes also merited consideration, Manitobans typically attributed decades of unsatisfactory drainage to government failings often involving suspected corruption. An even greater reliance on expert advice seemed to many to be a solution, and this contributed to a broad push for administrative reform in Manitoba.

Along with greater expert involvement in drainage came the introduction of the watershed concept. Defined in this context as an area of land united by a common drainage system, the concept made clear that Manitoba's drainage districts amounted to but small areas within much larger drainage systems. Particularly under the influence of the watershed idea, some government agents and flood-vulnerable landowners found it appropriate to spread drainage costs among all landowners in the watershed rather than among only those subject to flooding and included in a drainage district. But many Manitobans who farmed at higher elevations in watersheds that included flood-vulnerable areas were unwilling to contribute financially to solutions to what they perceived to be someone else's problem. The matter became the subject of intense public debate as involved Manitobans deployed concepts of private property and individualism in support of water management plans that favoured their interests. These competing colloquial liberalisms highlighted elevation as a significant factor in the creation of communities of interest in southern Manitoba.

Debate over financial liability for drainage works also revealed the contested nature of expert authority as farmers in higher areas disputed the expert recommendation that they should help to pay for drainage down below. Through documented public discussions and doubtless through countless private exchanges, interested Manitobans confronted not only contemporary drainage-funding proposals but also historical arrangements governing landholding in the province. Interested parties measured their personal convictions against the landscape of the province, sometimes reassessing their views of property and individualism in light of Manitoba's particular topography and water flows. In the face of enduring contention, the provincial government ultimately sought to appease all parties by undertaking (at government expense and under expert supervision) the construction of enormous drains intended to sever the

major connections between the upper and lower watersheds. This amounted to a dramatic new attempt to reconcile Manitoba's settlement system
and the local environment.


AFTER THE 1895 LAND DRAINAGE ACT

The early twentieth-century settlement boom across the Prairies coincided
with a drainage boom in southern Manitoba. The province's economy
expanded during this period, with a rise in government revenues fuelled
in part by the sale of lands that the federal government transferred to the
province as swamplands.[2] As a strong economy made further drainage
possible, agricultural expansion made it necessary. Eliminating wetlands
created more space for new settlers and allowed established farmers to
expand their operations, and reducing intermittent flooding enabled
vulnerable Manitobans to share in regional prosperity. By 1903, only eight
years after the passage of The Land Drainage Act, there were already
thirteen drainage projects under way.[3] They were numbered in order of
creation and identified by number. By 1914, eight more districts had been
added, encompassing slightly more than 2 million acres. Districts varied
significantly in shape and size, with the largest (District No. 2) including
nearly 450,000 acres and the smallest (District No. 13) just over 7,000.
Although three more relatively small districts were created in the mid- and
late 1920s (including District No. 24, which at 4,800 acres was even smaller
than District No. 13), bringing the total to twenty-four, the major period
of district creation was between 1896 and 1914.[4]

In large measure, drainage districts began as the ideas of government
experts. Engineers and surveyors employed in the Department of Public
Works had two important tasks. First, they were to propose district boundaries, encircling areas they deemed to be subject to flooding. These boundaries could be modified to accommodate public reaction, either enlarged
to include the lands of proponents or reduced to exclude the lands of
opponents in what has been termed a chiselling process.[5] Professional
authority was particularly important before 1913, when districts were proclaimed solely on the basis of expert recommendation, provided there was
not substantial opposition mounted by area residents. Although the need
for supporting documents endorsed by residents complicated matters after
1913, expert backing remained pivotal. Second, engineers were to identify
locations for drains intended to resolve flooding in the identified areas.
Throughout this process, extensive measurements were taken, and diagrams

FIGURE 10    Government survey party involved in drainage work, June 1911. Expert assessments were necessary to determine whether drainage was practicable. In areas where drainage seemed feasible, the survey party's measurements would be used in the creation of drainage district maps.
*Source:* Archives of Manitoba, Drainage 8, June 1911, Government Survey Party, Glencairn Camp.

reflecting drainage plans were drawn up.[6] If engineers doubted the viability of the proposed work, the project was at risk. For instance, despite enthusiasm by local residents, Drainage District No. 8 was delayed for years by the difficulty of conveying water through a substantial ridge of land separating the flooded lands from a safe outlet in Lake Manitoba.[7]

The process of creating drainage districts under the 1895 Land Drainage Act involved generating a new sort of knowledge about the wet prairie landscape. The documents produced through the creation of districts are strikingly different from the township plans created decades earlier by compiling the information that federal government surveyors collected. Some township plans provided a significant amount of environmental information. For example, major landscape features such as timber, swamp, and meadow were often depicted. They would be recorded as they were

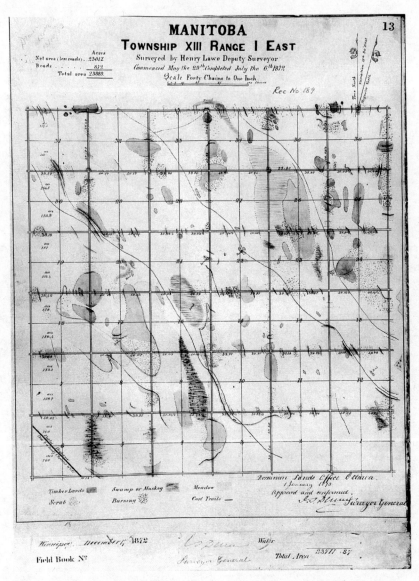

FIGURE 11    Township plan, township XIII, range 1 east. This is an example of the sort of document produced by the Canadian government through the work of dominion land surveyors.
*Source:* Archives of Manitoba, Township Plan, Survey Series, CN 177, 1872.

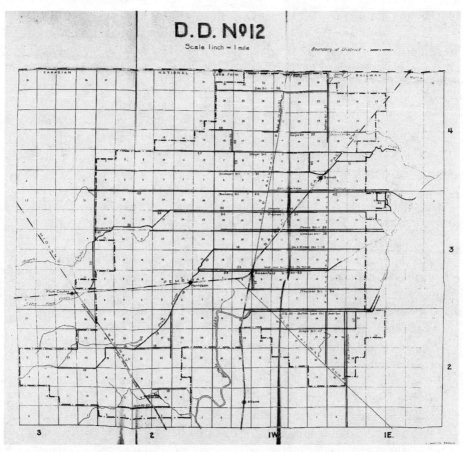

Figure 12    Map of Drainage District No. 12, prepared by the Reclamation Branch, Province of Manitoba, January 1931. The heavy (and sometimes double) lines are drains. The broken line is the boundary of the district, the small squares are sections within townships, and townships are numbered along the bottom and right side of the map. Rail lines and natural waterways are also evident.
*Source:* Archives of Manitoba, GR 43, G 61, file Drainage Districts, Drainage District No. 12.

found at the moment of survey, with little consideration of how they might vary from season to season or year to year. Other township plans included very little information about the landscape. The characteristic feature that remains consistent from township plan to township plan is the grid pattern dividing sections and quarter-sections. Surveyors laid out these lines and, if they chose, collected environmental information as they went, which explains why environmental information is often richer along the grid than within the squares. Township plans provided a snapshot view of the wet prairie, one framed by the township, section, and quarter-section divisions that defined land alienation on the Canadian Prairies.[8] In contrast, drainage plans depicted a flood-afflicted area that could be served by common drainage infrastructure. These maps encompassed multiple townships, reflecting underlying environmental patterns and important engineering considerations that extended across township boundaries. Juxtaposed with a township plan, a drainage district plan amounted to an argument for another way of conceptualizing the landscape, one oriented to environmental conditions instead of property boundaries.

The approach evident in drainage district maps was also a departure from the piecemeal efforts that had characterized much drainage work in Manitoba before 1895. In this early period, individual drains were the focus of planning and effort. Drainage district plans reflected attentiveness to the relation of each drain to the others and awareness of the entire area that the drainage system would affect. Government surveyors and engineers produced drainage districts by generating information that established the district as a meaningful environmental management entity. The digging of drains in drainage districts confirmed the legitimacy of these entities, anchoring them in Manitoba's prairie gumbo. Through this process, the communities of interest perceived by residents in flood-vulnerable areas took on greater scientific and political reality.

But the expert vision laid out in drainage district maps was not always realized on the ground. The reasons for failure were many, ranging from inappropriate design through changed environmental conditions to sabotage by residents unconvinced of the utility of the project. Drainage District No. 1 provides a well-documented instance of an unsuccessful drainage project. Located in a swampy area north of the forks of the Red and Assiniboine Rivers, the area had been a valuable source of wild hay for residents of the Red River settlement. By the late nineteenth century, however, surface water was inhibiting development in what was becoming a densely settled region. Under the Swamp Lands Act, the federal government transferred the inundated land to the province. Manitoba was eager to capitalize the lands that it received but recognized that it would be far easier to dispose of the lands once they were drained. A drainage district that encompassed provincially controlled swamplands as well as privately held homesteads was created. Since crown lands were not subject to taxation, the cost of drainage was divided among the private owners in the district, significantly inflating their costs. Inordinately high drainage taxes were part of what angered local residents. The more salient issue, however, was the fact that drainage was an utter failure.[9] The marsh was fed by a significant number of underground springs. Flooding in this region was not simply a question of runoff pooling in low areas. The springs ("subterranean rivers," according to an 1882 provincial report) meant that, regardless of alterations to drainage patterns and local topography, the area remained swampy.[10] Had the provincial government been as concerned with the environmental conditions of the area as with the potential for the province to profit from its drainage, the time, money, and effort invested in what proved to be a futile drainage effort might have been saved.[11]

Important among the various causes of ineffective drainage across Manitoba's wet prairie was the issue of patronage. From the earliest days of the province, governments were keen to use public works as a means of rewarding supporters, cultivating votes, or realizing personal profits. For instance, C.P. Brown, a prominent politician who spent many years as minister of public works, became involved in the Manitoba Drainage Company, an enterprise positioned to take advantage of the government drainage contracts available in the late 1870s and early 1880s. Although it appears that Brown extricated himself before the enterprise became profitable, the example suggests the connections between politics and business in the young province.[12] Over the years, there were a number of investigations into the role of patronage in Manitoba politics, many of which considered contracts for drainage.[13] Manitoba was hardly unique in this regard. At all political levels in Canada, government projects were a favoured device of political parties eager to improve their election chances.[14] The specific extent to which particular administrations or individual politicians participated in inappropriate practices matters little to a history of drainage in Manitoba. From an environmental history perspective, it is sufficient to appreciate that decisions about where, when, and how to build drains were affected by political considerations. Under both John Norquay (non-partisan, 1878-87) and Thomas Greenway (Liberal, 1888-1900), with no evidence of deviation through the brief administrations of David Harrison (non-partisan, 1887-88) and Hugh John Macdonald (Conservative, 1900), provincial governments intervened in local public works whenever it seemed advisable, for political and social reasons as well as in response to environmental conditions. In this way, the functioning of the Department of Public Works remained remarkably consistent through successive administrations, despite substantial political differences on other policy questions and despite The Land Drainage Act of 1895.

Nor did practices change under R.P. Roblin (Conservative, 1900-15), whose administration remains notorious for patronage. Indeed, in 1915, Roblin was obliged to resign in the wake of revelations about fraud and corruption in the construction of the new legislative building.[15] By the time of his resignation, the situation had become untenable, and a movement for government reform had coalesced. Manitoba electors favoured a change from Conservative to Liberal and brought T.C. Norris to power. The newly elected government launched an ambitious plan for change, one that included temperance legislation, women's suffrage, and credit programs for farmers.[16] In addition to these contentious issues, the new government also faced significant dissatisfaction with drainage. Some flood-vulnerable Manitobans were frustrated with government administration as they perceived that drainage efforts had been hampered by political considerations. For instance, in a 1917 letter to the Norris government, a drainage district resident created a compelling parallel to the scandal over construction of the legislative building that had rocked provincial politics: "There is a lot of us people around here [who] think this drainage scheme has been a fraud like the [Government] Buildings only on a smaller scale."[17] If the still-incomplete legislative building was an urban monument to patronage, some rural Manitobans saw something similar in the construction of ineffective drains. Both change in government and improvements in drainage were necessary.

Twenty years after the passage of The Land Drainage Act, many Manitobans still thought their agricultural production was hampered by surface water. This perception contributed to dissatisfaction with provincial governments, particularly as patronage was often suspected as a contributing factor. Manitobans had long been keen to take their drainage woes to the government. In addition to letters from individuals, there were petitions and collective submissions in which groups of various sizes aired their discontent. According to one government count, there were eighty-four separate petitions received over a thirty-three-year period in relation to Drainage District No. 2 alone.[18] Protest became more organized in 1918 with the formation of the Red River Valley Drainage and Improvement Association. Made up largely of representatives of the municipalities of Drainage Districts No. 2 and 12, with some participation from other municipalities dissatisfied with surface water management, the association lobbied the government for more effective action on drainage.[19] The government of Manitoba was receptive to the association's concerns and connected them to expressions of discontent received from individuals in the area and in other drainage districts. It was clear that something beyond Norris's increased public works funding was necessary.

The government responded in a manner in tune with its commitment to administrative reform, seeking expert advice on what had proved to be a difficult problem. Charles Gleason Elliot, a prominent American drainage engineer, was asked in 1918 to conduct an examination of the Manitoba drainage system.[20] Elliot served as chief of drainage investigations for the US Department of Agriculture and had published numerous works on drainage.[21] His *Engineering for Land Drainage: A Manual for the Reclamation of Lands Injured by Water* was in its third printing by 1919. Elliot was mandated to suggest the sort of inquiry that should be mounted into drainage problems in Manitoba and to offer preliminary thoughts on possible solutions. By evaluating all drainage districts in aggregate, Elliot was creating a context in which it was possible to perceive individual complaints as the local manifestation of a larger problem. Consulting Elliot was a means for the province to reaffirm the significance of technical expertise as well as a means to gain specific knowledge about the state of the province's drainage infrastructure. And involving an outside authority might have gone some distance toward appeasing Manitobans concerned that patronage might continue to weaken the drainage system.

Elliot submitted his report to the minister of public works in June 1918. Although general and non-committal in tone, it contained much that was favourable. Well acquainted with the challenges of drainage, Elliot appreciated the government's efforts to grapple with a difficult situation. Indeed, he thought the province was keeping pace with other regions of the British Commonwealth, many of which were in the process of "revising and perfecting their drainage systems and doing it at large expense."[22] An important part of this process of revision, in Elliot's view, was a public investigation into drainage. On his recommendation, the Manitoba Drainage Commission was formed by an order in council of 17 January 1919.[23] J.G. Sullivan, a civil engineer much experienced with railways, was appointed chair, and farmer H. Grills and entrepreneur J.A. Thompson rounded out the commission. The commissioners picked up where Elliot left off, assessing the province's drainage districts in aggregate and offering more detailed recommendations. They spent about two years examining the question of drainage through the study of local geography, research on drainage practices elsewhere, and consultations with experts and settlers. In December 1921, they submitted their report.

While the commission was at work, the provincial government was having a difficult time. Premier Norris had laid out ambitious plans for reform. The challenges of the years immediately following World War I, including a global economic depression and widespread labour unrest, limited what the government was able to achieve. Public dissatisfaction ensued, and in January 1922 a new provincial administration took over.

Although political involvement by western farmers had been on the rise for some time, the election of the United Farmers of Manitoba was still something of a surprise.[24] It was only after the election was won that the search began for a party leader who would become premier, and party members eventually settled on John Bracken. An agricultural scientist more than a natural politician, Bracken had just recently taken up the position as head of the Manitoba Agricultural College after a number of years working with the dominion experimental farms in Saskatchewan. Bracken enjoyed his work at the college, believing that careful scientific research into prairie agriculture would improve the lives of prairie farmers. Ultimately, it was his personal commitment to working on behalf of agriculturalists that led him to accept the job as premier.

The Manitoba Drainage Commission submitted its report just weeks before the Bracken administration came to power. Not surprisingly considering his scientific background, Bracken was receptive to the contributions of experts. He also advocated the related cause of business-like administration, defined in opposition to the party politics that had coloured many aspects of provincial administration in the past. The investigation into drainage launched by Norris had proceeded along lines in tune with Bracken's own beliefs. For instance, his supporters were, if anything, equally committed to non-partisan administration. At hearings before the Legislative Committee on Drainage in the early 1920s, the suggestion that government committee members might be unduly biased in favour of certain projects provoked a raucous dispute.[25] Government officials reacted strongly because of their public commitment to government reform. They were determined not to be seen to replicate the patronage activities of past administrations, which had hindered the development of the drainage system. Although the commissioners likely regretted the fall of the Norris administration that had appointed them, there was ample reason to anticipate that the new government would take seriously their recommendations.

Historian Doug Owram, in his important history of Canadian intellectuals and the state in the first half of the twentieth century, argues that "only slowly and with some difficulty did the areas of science and engineering establish themselves within the public service as areas of professional expertise."[26] Owram identifies a period of transition between 1918 and 1929, from the patronage politics of an earlier era to the bureaucratic administration that was becoming associated with progress.[27] At least in rough terms, Manitoba fits this pattern. As historian W.L. Morton noted decades ago, the 1915 downfall of the Roblin administration and the reforms that followed "marked the end of the crude politics of frontier days in Manitoba."[28] Although in Owram's vision increased reliance on technical expertise was promoted by urban reformers, drainage in southern Manitoba illustrates how the impetus for change also grew out of problems in the rural context.[29] This is particularly important in assessing the process of government reform on the Prairies, for the region's population remained substantially rural far later than that of eastern Canada. There might be a general correlation between an urban reform movement and the rise of technical expertise in many parts of Canada, but in Manitoba at least, dissatisfaction with rural drainage was an important impetus for change in government practices.

## DRAINAGE DISTRICTS AND THE WATERSHED

Manitobans hoped that correcting the administrative problems that be-devilled drainage schemes would solve the enduring problems of surface water flooding. The experts who assessed Manitoba's drainage system in the late 1910s and early 1920s certainly underlined the importance of objective administration. But still they recognized that relatively localized deviations from systematic drainage (whether due to patronage in government decision making regarding drainage or any other cause) were not to blame for the major physical problem with the drainage infrastructure. At the root of much inadequate drainage was a basic difficulty derived from the mismatch between the conceptualization of drainage districts and the topography of the province. Although drainage districts were defined in relation to water flow patterns, they were limited to lands that were subject to substantial and frequent inundation.[30] Ultimately, drainage districts encompassed only a fraction of the watersheds in which they were situated. In both C.G. Elliot's 1918 analysis and the Manitoba Drainage Commission's 1921 report, it was argued that Manitoba's drainage districts should be expanded to reflect their watersheds.

The principle of management by watershed was gaining favour internationally, but it was provincial topography that made it particularly important. In the soup bowl of southern Manitoba, the comparatively flat basin of the Red River valley was bounded on the east by the irregular topography of the Precambrian Shield and on the west by the dramatic increase in elevation along the Manitoba Escarpment.[31] The watersheds of

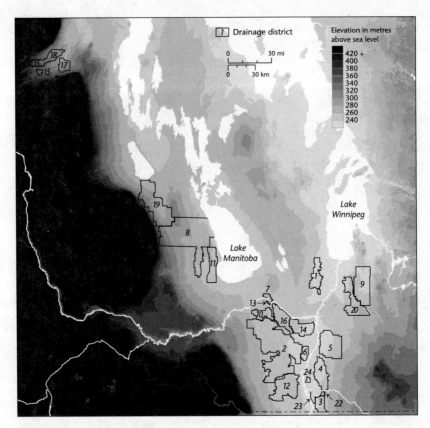

FIGURE 13     Southern Manitoba showing elevation of drainage districts
*Sources:* Adapted from Archives of Manitoba, Manitoba Department of Mines and Natural Resources, Surveys Branch, Map of Manitoba, Southern Portion, 1933; the Drainage Districts, map, in J.H. Ellis, *The Soils of Manitoba* (Winnipeg, Manitoba: Economic Survey Board, 1938), 29.

the province extended from the lowlands to the uplands, yet most of the drainage districts of the province were limited to the lands at or near the bottom of the Manitoba soup bowl.

What did this mean for drainage? In 1950, provincial government investigator R.H. Clark described the situation in the following terms:

The tributaries which originate in the rolling hills to the west and the high flat marsh plains to the east have steep slopes of entrance to the valley. Upon reaching the ancient lake bed [the bed of glacial Lake Agassiz] these slopes quickly flatten out and, since the channel through the plain has not sufficient capacity to carry the flows, the water quickly tops the banks and spreads out over the valley.[32]

In the evocative language employed by early drainers, watercourses would "lose themselves" as they passed from the sides to the bottom of the soup bowl. Water would spread out across extensive areas of land. Some of it eventually found its way to the lowland waterways, but much of it pooled in slight depressions and disappeared only through evaporation.[33] Although some areas were afflicted more severely than others, and some districts confronted more localized challenges to drainage, the problem of water pooling in the bottom of the bowl was common to many areas of southern Manitoba. In the view of international expert C.G. Elliot, the problem was "peculiar to this region," at least in its severity.[34] The original layout of Manitoba drainage districts reflected a spatially restricted conception of surface water problems. There was no effort to accommodate the overland flow that contributed to flooding.

Expert recognition of the problem merely confirmed what many lowland Manitobans already recognized. With drainage districts defined by the spatial extent of flooding, rather than in relation to the larger watershed in which the flooded lands were embedded, the water that ran in from higher lands became known as foreign water. The distinction between foreign and local water was central to conflict over drainage in Manitoba. Rather than a proliferation of limited upstream/downstream conflicts among individuals, relatively coherent geographically defined communities of interest emerged. In both the uplands and the lowlands, interpersonal conflicts related to surface water were dwarfed by the general significance of one large slope of consequence. Foreign water was less a matter between neighbours and more a matter between localities differentiated by elevation. Because of this, individuals' feelings of anger and resentment coalesced through discussions with others who were similarly afflicted. The terms "highlanders" and "lowlanders" became part of the local discourse, consolidating communities of interest among those who occupied similar topographical locations and suggesting conflicts of interest with those who lived at different elevations.[35]

These communities of interest espoused relatively coherent doctrines that amounted to colloquial liberalisms: abstract liberal principles as understood and deployed by non-experts in the context of a particular social, political, and environmental landscape. In Manitoba's wet prairie, colloquial liberalisms shared some elements with more formalized liberal doctrine. For many Manitobans, for instance, the private property ideal provided the liberal pivot around which they spun their arguments, with lowlanders claiming that their ability to use their land was hampered by highland water and highlanders asserting their right to use their land without concern for any lowland consequences. In justifying their claims, many Manitobans invoked their personal experiences on the land. These were deeply felt arguments rooted not in intellectual abstractions but in the particular geography of an individual farm. Two conflicting strands of colloquial liberalism had emerged, each with its own geographically based view of what was fair.

To better understand the development and operation of these communities of interest, it is helpful to focus on a particular location. Drainage District No. 2, situated between the Manitoba Escarpment and the Red River south of the Assiniboine River, had a severe problem with foreign water. The Manitoba Escarpment is a key topographical feature of the province – the highest and steepest side of the soup bowl. In an examination of the escarpment, geographer W. J. Carlyle provided a good description of the environmental factors that contributed to the foreign water problem in Drainage District No. 2. In the spring months, the flow of meltwater from the escarpment was often delayed by the snow and ice that persisted on the lowlands. Although precipitation was generally lower to the west, the highlands were vulnerable to intense rainstorms in May, June, and July. The swift-moving waters carried a great deal of sediment eroded from the shaly till and shale bedrock of the escarpment.[36] Annual snowmelt flooding that overwhelmed lowland streams, flash floods caused by intense precipitation, and the sediment that clogged waterways and ditches: all clearly descended from the highlands to the west.

Basic physical geography was not in itself a catalyst to dispute. Most lowlanders were content to tolerate some foreign water, believing that the lowlands should absorb what was understood as the natural flow from the highlands. Drainage District No. 2 was incorporated in 1898, and work according to the original plan took until 1907 to complete. Through a series of main and feeder drains branching out across the area, flow patterns were fundamentally reshaped. The area had included a few substantial wetland complexes such as the Boyne Marsh and the Tobacco Creek Marsh. Drainage efforts zeroed in on these locations, and ditches were constructed in number and capacity appropriate to the existing problem. The massive project was buoyed throughout by the optimism of experts and residents who anticipated a reprieve from flooding.

At the same time, however, rapid environmental change was also taking place beyond Drainage District No. 2. Settlement, clearing, and plowing had increased in areas both within and beyond the drainage district. The 500,000 acres (202,000 hectares) of Drainage District No. 2 accounted for barely a third of a watershed of over 1.2 million acres (nearly 500,000 hectares).[37] Although flooding in the district was moderated for a few years by dry conditions, 1912 brought severe inundation. This inaugurated another period of intense construction that concluded about 1915. By this point, frustrated with continued surface water problems, lowlanders were beginning to look beyond the drainage district to explain flooding. In their opinion, the flow regime had been changed. They began to argue that ongoing cultivation and deforestation in the longer-settled regions along the escarpment were affecting flow patterns in ways detrimental to the lowlands. As lowland farmer L.C. Wilkin explained in 1922, he and others in a comparable situation recognized that they were "not facing a natural condition by any means." Rather, their farms "have been made a dumping ground for all the water from a large area."[38] Lowlanders thought that agricultural progress in the lowlands was hampered by the changed flow patterns derived from agricultural progress in the highlands. Foreign water no longer seemed to be entirely natural.

The Sullivan report explained that "the greatest factor causing damage from flooding is the changed conditions since the districts were first formed."[39] Drainage District No. 2 and its watershed provide an example. In seven townships just beyond the western boundary of the district, there were 4,177 acres (1,690 hectares) of improved land in 1877. Some thirteen years later, the amount of cultivation had increased by nearly 4,000 acres (1,618 hectares). In 1915, there were 82,000 acres (33,184 hectares) improved. By the early 1920s, nearly all of the land in the area had been cleared of timber and planted to crops. In lands farther west but still within the watershed, commissioners put the rate of cultivation at 75 percent. The problem, as the report explained, was that such changes worsened flooding in the drainage district: "No matter how well the original channels were designed, and no matter how well they were maintained, they would not now be capable of properly taking off the extra rush of waters that comes from the higher grounds on account of the changed conditions."[40] The expert perspective supplied by the Sullivan Commission paralleled the lowlander viewpoint. Highlanders should help to fund lowlander drainage in proportion to the extent to which land change on the highlands had increased flooding on the lowlands. According to the commission, the first step

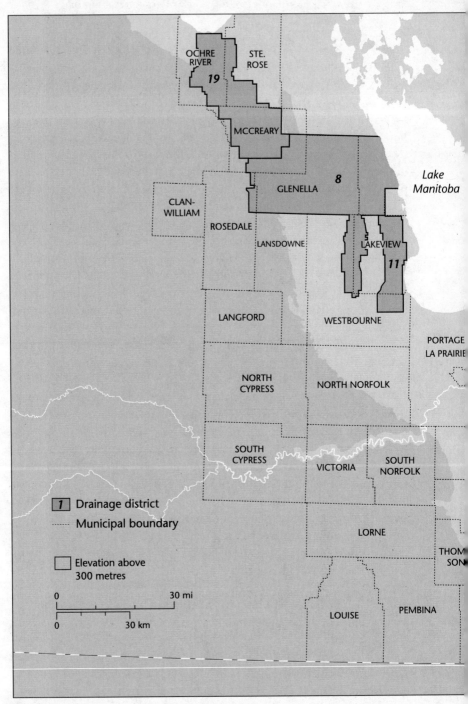

FIGURE 14 Municipalities and drainage districts, 1933

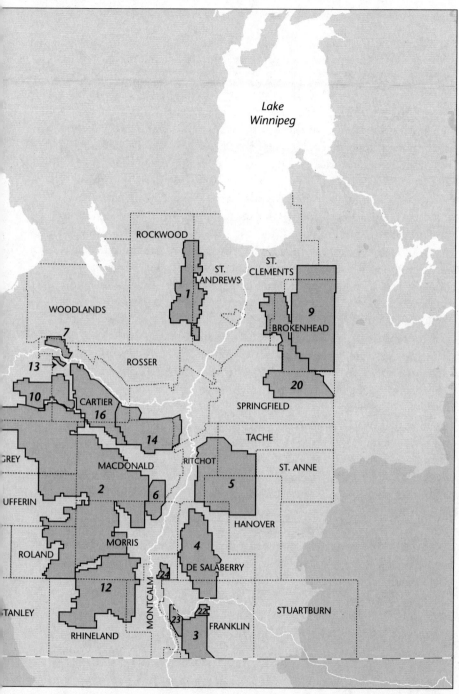

Lake
Winnipeg

ROCKWOOD

ST.
ANDREWS

ST.
CLEMENTS

*1*

*9*

BROKENHEAD

WOODLANDS

*7*

*13* →

ROSSER

*10*

CARTIER

*16*

*20*

SPRINGFIELD

*14*

TACHE

GREY

MACDONALD

RITCHOT

ST. ANNE

*2*

*6*

*5*

UFFERIN

HANOVER

MORRIS

*4*

ROLAND

DE SALABERRY

*12*

*24*

MONTCALM

*22*

STUARTBURN

TANLEY

*23*

FRANKLIN

RHINELAND

*3*

*Sources:* Adapted from Archives of Manitoba, Manitoba Department of Mines and Natural
Resources, Surveys Branch, Map of Manitoba, Southern Portion, 1933; the Drainage Districts,
maps, in J.H. Ellis, *The Soils of Manitoba* (Winnipeg: Economic Survey Board, 1938), 29.

would involve expanding the boundaries of the drainage district to "include all lands whose surplus waters drain into said district and are carried by an artificial channel through it to a natural outlet."[41] With drainage districts expanded to approximate more closely the watershed, highlanders could be made to contribute to the costs of draining the lowlands.

Following submission of the report in December 1921, and reflecting the continued importance of public engagement on drainage questions, the Legislative Committee on Drainage convened a series of public hearings. Partial transcripts of the proceedings have survived and provide a vivid picture of how the watershed idea was received.[42] Not surprisingly, opinion often broke along topographical lines. The province had sent notice of the hearings to the municipalities, and those who testified were identified as representatives of specific municipalities. Many from municipalities that overlapped substantially with one or more drainage districts were sympathetic to the idea of watershed management. For instance, lowlander McCallum from the Rural Municipality of Roland, the eastern half of which was included in Drainage District No. 2, argued that "there seemed to be no other logical boundary for [drainage] except the watershed."[43]

Highlanders were unconvinced. They contested the relationship between action in one region and effect in another that underpinned the idea of watershed management. Reeve D.F. Stewart of the Rural Municipality of Thompson, situated directly west of the Rural Municipality of Roland, argued that the Sullivan report was "entirely wrong" in its assertion that extensive cultivation had led to hastened and increased runoff. He was blunt: "Any farmer who knows anything about farming knows that cultivated land will not run off water faster than wild land." There was a logic behind his assertions: "The frost is out two feet on the cultivated land before it is out of the grass land, and it is free therefore to absorb more water." According to Stewart, commissioners had an erroneous understanding of the environment. He sought to enlighten them, explaining that in many parts of Thompson "we are holding that water back all we can, and would like to get more. We have lost our crops for want of water. It is not drainage but irrigation we want."[44] Thompson is an interesting case since the municipality lobbied the government for action on surface water management through membership in the Red River Valley Drainage and Improvement Association but objected to watershed-based drainage funding as proposed by the Sullivan Commission. The Rural Municipality of Dufferin, which included a small portion of Drainage District No. 2, also participated in lobby efforts but opposed Sullivan's recommendations. Perhaps this fracture within the association explains why it does not seem to have endured much into the 1920s. Dissatisfaction with surface water management did not necessarily amount to support for watershed management as recommended by the Sullivan Commission.

Sullivan had not anticipated that he would encounter serious objections. Indeed, he was entirely confident that all would see the inherent justice of aligning drainage district boundaries with the watershed and assessing all landowners within the watershed for a portion of the expense of drain construction.[45] His confidence was reinforced by officials in the Manitoba Department of Public Works. H.A. Bowman, chief engineer in the Reclamation Branch, wrote to the chair of the Legislative Committee on Drainage in 1922, explaining that, after having "given this matter considerable study," he had come to the conclusion that management by watershed "is the only solution to the acute problem before us." Furthermore, Bowman was sure that it would "be hard to find any Engineer of repute who has made a study of reclamation who will hold a contrary opinion."[46] However, scientific consensus on watershed management was not complete. The experts cited by the Sullivan Commission emphasized that, as a general rule, tree cutting and crop planting increased both the speed and the amount of runoff. But the link between action in one part of the watershed and effect in another remained diffuse and controversial. Indeed, engineer Bowman himself noted in a 1929 letter that "opinions are largely divided among Reclamation Engineers, the majority, perhaps, being of opinion that growth or denudation of timber has little or no effect on total runoff."[47] Contrary to Sullivan's expectations, even experts disagreed over the effects of highland conditions on the lowlands.

Highlanders were not mollified by Sullivan's assertions that they would be liable only if improvements to highland areas had altered surface water flow and that their assessments would in any case be comparatively moderate. To say that highlanders rejected the idea of apportioning drainage costs on a watershed basis does not adequately convey the fury of their responses. Highland farmer Black, representing the Rural Municipalities of Thompson and Stanley, found it "incredible that any legislative body should attempt to put across such a despotic and arbitrary matter as this."[48] To highlander Goldring from the Rural Municipality of Victoria, the report of the Sullivan Commission seemed to be a "diabolical" plot.[49] And Reeve D.F. Stewart from the Rural Municipality of Thompson asserted that those opposed to the report would fight against it "until the last ditch."[50]

Why such strident language? For highlanders, the recommendations of the Sullivan Commission were a fundamental betrayal of the terms under which they had taken up land in the province. This issue was at least as important as the accuracy of the science Sullivan invoked in support of his recommendations. Highlanders thought that they were being told they had contributed to lowland flooding not by wrongdoing such as irresponsible drainage but by the apparent good work of clearing and cultivating. Under the Dominion Lands Act, homesteaders were required to make improvements to secure title to their lands. Those who purchased land did the same to capitalize on their investment. Highlanders who cleared land had been doing just as they had been encouraged – even obliged – to do by government. But the Sullivan Commission now asserted that, by improving their land, highlanders had rendered themselves liable for damage to the lowlands. How could this be?

Beyond this question, there was the issue of whether settlers had any responsibility for quarter-sections other than their own. Much highlander testimony before the Legislative Committee on Drainage consisted of settlement experiences on quarter-sections with challenges every bit as daunting as the flooding that affected lowlanders. In the end, the complex and diverse experiences of highland settlers were reduced to one substantive point. Since they had received no assistance with the hard tasks and significant expenses they faced on arrival, why should they assist with drainage? J. Rose, a highlander from the Rural Municipality of Ochre River (which included lowland drainage districts as well as highlands), explained how, on arriving in the region, he had paid $4,500 for a highland quarter-section that had already been cleared of timber. His brother had paid $1,700 for undrained lowland that was subject to flooding.[51] In his mind, the difference in their initial outlays represented the price of improvement, whether what was needed was clearing or drainage. Having paid the higher price, he was unwilling to subsidize his brother. Not even familial relationships moderated the conviction that individuals should fund improvements to their own homesteads. The situation was exacerbated since lowland farms were generally more productive than those in the highlands.[52] Highlanders particularly resented the prospect of having to contribute to the reclamation of "land which is worth much more than ours," in the words of Reeve Stewart.[53]

Before the Legislative Committee on Drainage, highlanders invoked the fundamental legislative and ideological pillars that underpinned the settlement of the Canadian Prairies. The Dominion Lands Act linked private property and agricultural progress as homesteaders proved up by cultivating their lands. The colloquial liberalisms espoused by many Manitobans reflect this linkage. In southern Manitoba's soup bowl, however, progress for some meant disaster for others. Watershed management seemed to lowlanders to be a solution, but to highlanders it seemed to be a betrayal. The watershed concept would have redefined property and progress by establishing a relationship between the upper and lower watersheds. Since highlanders could have been obliged to contribute to the cost of lowland drainage, they would have been involved in the improvement of lands they did not own. Not surprisingly, they objected. The watershed idea did not accommodate the human history of the landscape – the notion of private property legitimized by agricultural improvement that remained, at least for highland settlers, entirely naturalized. It was clear that the province's drainage problems would not be easily solved, but neither would the human landscape of progress and property be easily displaced.

Addressing conflict in Idaho's irrigated landscape, environmental historian Mark Fiege emphasized how distance from the water source affected irrigators' views of how water should be managed.[54] Similarly, the conflicts in Manitoba were related to location on the landscape. Significantly, location could also become a catalyst to cooperation. As Wilkin explained before the Legislative Committee on Drainage in February 1922, he and his fellow lowland farmer Kirk appeared not "as individuals" but as representatives of the lowlanders, who, in recognition of shared concerns, had "clubbed together" to present their case more forcefully.[55] Examining the reception of the watershed idea makes clear that environmental factors contributed to the creation of meaningful group identities, ones rooted in differences of elevation (highlanders or lowlanders) but manifest through views on basic land management principles (private property as an absolute or as moderated in relation to the ecological commons of the watershed). These communities of interest reflect neither the ethnic nor the class divisions that have been emphasized by many prairie scholars.[56] They are not reducible to political loyalties, despite the enduring importance for drainage of the relationship between politicians and their constituents. They amount to deeply felt expressions of liberal principles as understood by non-experts in relation to their particular environmental experiences.

THE ECOLOGICAL COMMONS AND CREATIVE THINKING

By the time of the Sullivan Commission, agricultural cooperation was well developed across the Canadian Prairies. The Grain Growers Association of Manitoba, an outgrowth of the Territorial Grain Growers Association, was organized in 1903.[57] By the late 1910s, farmer parties were experiencing success provincially and nationally.[58] These developments signalled the emergence of prairie farmers as a political and economic force for improvement in the lives of rural dwellers. Disputes between highlanders and lowlanders took place at roughly the same time as many farmers were working together to achieve common economic and political ends. Ultimately, the transcripts of the debate over the Sullivan report illustrate that the development of a farmers' movement did not prevent disputes among farmers who perceived their interests to be in conflict. In the early 1920s, highlanders did not see how watershed management could possibly be to their benefit. Although it might increase the agricultural yield from the watershed as a whole, it would likely mean, in a landscape divided into privately worked quarter-sections, an expense to some and a benefit to others. If prairie cooperative movements were intended partly to redress the exploitation of prairie producers by outsiders, watershed management was a means to redress what lowlanders saw as exploitation by highlanders. The controversy before the Legislative Committee on Drainage was distinguished from the regional and national issues in which farmers intervened in part by how watershed management concerned inequality within the farmer community rather than injustice perpetrated by outsiders.

The private property ideal was, for many on the Prairies, a source of pride and an element of shared identity. Yet prairie settlement depended heavily on shared infrastructure construction undertaken in the liberal tradition, such as the road system established and maintained by provincial and municipal governments. As numerous historians have recognized, the individualism associated with prairie farming diverged from the manner in which prairie settlement actually proceeded.[59] At the public hearings hosted by the Legislative Committee on Drainage, the private property ideal provided a rhetorical recourse to highland farmers who sought to construct convincing arguments against watershed-based drainage financing. Such absolutist thinking belied the liberal compromises that underpinned life in the prairie region. Ultimately, highlander use of the private property ideal starkly conflicted with the realities of regional development.

But even highlanders, outraged though they were at any suggestion of financial liability for lowland drainage, found themselves considering other ways of thinking about private property and public infrastructure. In Manitoba's wet prairie, the state of the local environment and the character of necessary public works prompted some Manitobans to reassess the colloquial liberalisms they had espoused. It was the question of road building in a flood-prone area that provided the catalyst. Extensive and thorough drainage was of course necessary to ensure decent road conditions through-out many parts of the province. As municipal representatives and government officials considered the agricultural consequences of differences in elevation, roadside ditches were identified as a means by which improvements in the highlands had a detrimental effect on the lowlands. Leaving aside for the moment their unwillingness to accept any connection between actions on the highlands and effects on the lowlands, highlanders countered with a different sort of argument. It was submitted that anyone who travelled through the highlands, not only residents, benefited from road improvements. Indeed, it was asserted, provincial prosperity was tied to the existence of a road network. Representatives from highland municipalities argued that, because all Manitobans had an interest in road building, the entire province should be liable for any damage caused by roadside drainage. Highlanders alone should not be obliged to compensate lowlanders for increased flow resulting from road construction. The whole province should pay.

This line of reasoning quickly led to complexity. If the province at large had an interest in roads, perhaps the province at large had an interest in drains. Drainage was necessary for agricultural progress. Agriculture was fundamental to the future of the province. Therefore, every Manitoban was interested to the extent that all would benefit from the provincial prosperity an effective drainage system would support. Not surprisingly, representatives from the lowlands were sympathetic to this perspective. Some argued that the entire province should be considered a single drainage district, with the result that the cost of drainage would be shared among all Manitobans.[60] It was not only victims of flooding who found this a persuasive line of argument. Reeve Morten of the Rural Municipality of Westbourne, who had never been subject to flooding and who represented an area that included a relatively small amount of drainage district land, argued that the cost of drainage should be distributed among all Manitobans. "I am in favour of taking the whole province into the drainage district," he submitted; "make it wide."[61]

But the argument for provincial funding of drainage seemed tenuous to some. Although all Manitobans would benefit from improved roads, so would visitors from Toronto.[62] If agricultural production was the concern, then drainage could have consequences that transcended even national borders. An article describing drainage in Manitoba published in the *Canadian Engineer* in the late 1910s identified "thorough and systematic drainage" as "the only hope of insuring good crops from year to year."

Good crops were essential if Manitoba were to fulfill its role in the empire. As explained in the article, "the basic industry of Manitoba is agriculture and the greatest economic service we can render the Empire is to increase the production of foodstuffs."[63] Particularly in the years following World War I, Manitoba's imperial role was no minor consideration. Clearly, determining who had a stake in drainage was not an easy task. Beyond the perhaps too simple arrangement that saw individuals solely responsible for improvements to their own homesteads lay a world of complication. There was no natural or obvious scale (neither a homestead nor a watershed nor a political unit such as the province) at which projects should automatically be financed.

The complexity derived in large part from the fact that this was ultimately a question of values.[64] Would Manitoba be a place where private property prevailed absolutely, with farmers free to disregard how their actions affected others? Or would it be a place where land use would be coordinated across individuals' holdings, with highlanders compensating for how their actions might negatively affect lowlanders? At stake was the human as much as the physical landscape of the province. Understanding this allows a fuller appreciation of the significance of the drainage debates. Defined in terms of the identities of participants, debates over drainage in the first third of the twentieth century were largely local. Despite the importance of occasional interventions from international experts such as C.G. Elliot, contributors were overwhelmingly Manitoban: rural landowners, local professionals, and municipal and provincial governments. Nevertheless, the terms of debate were certainly not provincial. Working out the matter of drainage meant considering broader questions. What was the interest of the province, the nation, the empire in drainage? What was the appropriate relation of settlers to their land and of Manitobans to each other? Conditions in Manitoba's wet prairie catalyzed debate over the basic legislative and ideological principles governing landholding across the prairie region. This was liberalism cast into relief as the troublesome geography of Manitoba's wet prairie prompted some to reflect anew on such matters.

Despite the importance of the distinction between highlander and lowlander perspectives, divisions faded somewhat as all grappled with the question of how best to make use of a challenging environment. Even in the context of public hearings that heightened antagonism by providing a forum for confrontation, many Manitobans displayed a capacity for creative thinking about how best to live in the region. This dialogue encompassed the possibility of changed landholding practices, a radical departure from one of the basic elements of agricultural life in the province. And, though debates over drainage certainly drew on a wider cultural context, including the growing influence of international experts and an internationally meaningful private property ideal, these deliberations were catalyzed and defined in large measure by environmental conditions in southern Manitoba's wet prairie. Water flow between the highlands and the lowlands remained controversial, but the matter of drainage provided a link between the seemingly disparate scales of broad ideological principles and local environmental conditions. At least in Manitoba's wet prairie, they flowed together.

Creative thinking by Manitobans did not inspire concrete action by the provincial government. The key recommendations made by C.G. Elliot and the Sullivan Commission were allowed to languish by Bracken's farmer government. Two key signposts of this administration – fiscal economy and expert authority – pointed in opposite directions. Bracken was actively shrinking the civil service at a time when Sullivan recommended the establishment of a permanent commission to govern drainage.[65] Although the commission was established, it was always under government pressure to minimize costs.[66] After a year and a half, Sullivan resigned from the position he had been granted on the permanent commission, believing that he was not providing value for money to a government dedicated to efficiency.[67] Politics were also a major consideration. As drainage brought farmers in conflict with each other, it split the political constituency to which Bracken appealed. Among the province's staunchest Brackenites were those of British origin, and they made up a larger proportion of the population in the highland than the lowland municipalities.[68] Despite his desire to govern according to the sort of expert guidance offered by the Sullivan Commission, Bracken had an interest in maintaining the goodwill of this group. And the scientific uncertainty that came out before the legislative hearings might well have made it easier to brush aside Sullivan's recommendations.[69] For Bracken and others in his government, inaction was the option most likely to preserve the support of both highlanders and lowlanders.

Among the few Sullivan Commission recommendations that produced government action was the construction of double dyke drains. A double dyke drain consisted of two ditches excavated at some distance from each other. The material taken out of each drain was deposited on the bank farthest from the other drain, creating levees that enclosed the two ditches as well as the expanse of land between them (see Figure 15).[70] In years of regular flow, water would be confined to the two drains. During freshet, the levees served to contain the early spring flow that often occurred before

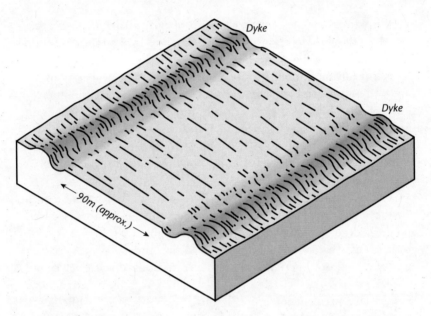

FIGURE 15    Double dyke drain
*Source:* Adapted from John Warkentin, "Water and Adaptive Strategies in Settling the Canadian West," *Manitoba Historical Society Transactions* 3 (1971-72): 72.

the ice had cleared from the drains. In times of extreme flooding or heavy freshet, the levees provided an additional measure of protection for the surrounding area as water could flow over the land between the drains. Early on, double dykes represented a means of coping with extreme flows. The Sullivan Commission's recommendation led to the construction of the Norquay Channel, the 4N Drain, and the Tobacco Creek Channel. Construction costs were borne by the drainage districts they traversed, and the first of the double dykes was completed in 1923.[71] Over subsequent years, more double dyke drains were constructed. Particularly significant were the double dykes along the Elm Creek Channel and the Shannon Creek Channel in Drainage District No. 2 and the Hespeler Channel and Rosenheim Channel in Drainage District No. 12.[72]

Double dyke drains remained an important part of government drainage strategy through the policy adjustments of the 1930s, which focused largely on the issue of maintenance. Despite the significance of the changes of these years, which reflected a new understanding of the need to care for the drainage infrastructure, the boundaries of the districts themselves were not substantially altered. As a result, the mismatch between the drainage district and the watershed went unresolved. Lowlanders were still incensed at how land use changes on the highlands created flooding that they had to pay to address. Highlanders were no more willing to contribute directly to the costs of draining foreign water. Double dyke drains dealt with the technical aspects of surface water drainage, providing a more effective means of controlling extreme flows. But a construction technique alone did not address the complicated social question of who should pay for foreign water drainage.

Before the Legislative Committee on Drainage, much debate was polarized between highlanders and lowlanders. As we have seen, there was also some comparatively open-minded consideration of alternative financing arrangements, catalyzed through discussion of the various interests that would be served through resulting improvements to routes of travel or agricultural production. Beyond both of these utterly entrenched positions and this more philosophical approach were the pragmatic concerns that preoccupied Manitoba civil engineer M.A. Lyons. The early 1940s had seen a rise in dissatisfaction with drainage prompted largely by higher rates of precipitation. Financial losses attributable to drainage were piling up. In the area along the Red River south of Winnipeg, the reduction in agricultural income in 1943 due to flooding was $3,150,000. In 1944, it was $10,900,000. The situation improved slightly in 1945, though the number was still staggering: agricultural incomes that year were reduced by $5,200,000.[73] In 1947, the government responded by launching an inquiry. Recently retired from the Department of Public Works after many years spent administering Manitoba's flawed drainage system, Lyons was ideally suited to conduct the investigation.[74]

Among his key concerns were the logistics of watershed-based drainage financing. Even if it was agreed that lowlanders should not be obliged to pay for foreign water drainage, how would this work out in a practical way? Distinguishing foreign water from local water within a lowland drain was no simpler than measuring the downstream consequences of upstream land use changes. Would it even be possible to establish how much highlanders should pay, given the difficulty of determining what proportion of flooding was attributable to foreign water? Lyons recognized the problem and saw in double dyke drains a potential solution. He argued that, if constructed in sufficient numbers and carefully chosen locations, they could maintain a physical separation between highland and lowland water. Having accepted that it was likely impossible to achieve consensus on watershed management, he began advocating that the best solution was to attempt to sever the major links between the upper and lower watersheds. Smaller channels in the highlands would funnel surface flows toward the double dyke drains, which would then speed the water through the low-lands to safe outlets in natural rivers or streams. The levees along the outer banks of double dyke drains that prevented unusually large flows from spilling over onto the lowlands would also keep the drains from collecting any lowland water. The result would be largely separate systems for the management of highland and lowland water. In this situation, it would be far easier to obtain the environmental equivalent of separate cheques for highlanders and lowlanders.

This use of double dyke drains was entirely different from the approach advocated by the Sullivan Commission. The commission had argued that drainage districts should be expanded to approximate more closely the watershed since doing so would allow the expense of drainage to be spread across all lands that contributed to flooding. In contrast, Lyons proposed to preserve the viability of something like the drainage district as originally conceived. Instead of expanding districts to approximate more closely the watershed, patterns of water flow would be altered to conform to estab-lished district boundaries. The watershed was to be remade to accommodate the drainage district. This change of approach was partly a question of era: by the 1940s, engineering projects were often designed to work with nature. But it was also a question of lessons learned: decades of dispute had made apparent that some compromise was necessary.

In his 1921 report, Sullivan argued that drainage district boundaries should reflect the natural boundaries of the watershed in which they were situated, at least as far as possible. Whereas Sullivan was embracing an appealing abstraction, Lyons was grappling with both a problematic physical terrain and a conflict-ridden human environment. Concerned with which management arrangements were humanly possible, Lyons was obliged to rein in the environmental idealism that pervades the Sullivan report. In effect, Lyons recommended that surface water patterns be remade to reflect the erroneous assumptions of government administrators and early settlers who, with little appreciation for the ecological commons of surface water, assumed that individual homesteads could be owned as private property and managed in isolation.

Given his pragmatic approach, Lyons could not avoid the fact that the simple availability of separate cheques for highlanders and lowlanders failed to resolve the difficult question of who should foot the bill for the drains to manage water running off the highlands. An investigation under-taken a decade earlier had suggested the beginning of an answer to this difficult question. Launched by The Land Drainage Arrangement Act of 6 April 1935, the second royal commission investigating drainage in

Manitoba was chaired by Professor John N. Finlayson of the Faculty of Engineering at the University of Manitoba, and it included John Holland, reeve of the Rural Municipality of Springfield, and John Spalding, secretary of the Union of Manitoba Municipalities.[75] When the commission reported in March 1936, it recommended that the province assume about 45 percent of the accumulated debt of the drainage districts.[76] A more substantial flow of cash from the province was, in Finlayson's view, the best way to establish districts on a firmer footing and help to reduce conflict over flows of water from highlands to lowlands.

Although Finlayson proposed that the government undertake certain improvements immediately and make an annual contribution to the cost of maintenance, municipalities were informed in May 1936 that, because of financial difficulties associated with the Great Depression, the government was "unable to undertake the capital expenditures involved in some of these recommendations."[77] Both economic circumstances and perceptions of what constituted appropriate government undertakings changed dramatically through the years of World War II,[78] and in 1945 Minister of Public Works E.F. Willis doubted "whether the Provincial Treasury pays a fair share of the cost of the work within the Drainage Districts in this Province."[79] Building on both the earlier recommendations of Finlayson and on the growing political willingness expressed in the comment by Willis, Lyons recommended that the double dyke drains that were to manage highland water be funded by the province at large. He recommended something similar to what had been suggested by Westbourne Reeve Morten twenty years before: to include the entire province in a single district for the purpose of financing drainage.

Lyons also justified his recommendation for more substantial provincial funding with reference to the benefits that would accrue to the province through improved drainage in needful areas. Explaining his recommendation that the province assume more of the responsibility for funding drainage, Lyons pointed to what he saw as a fundamental principle of the 1895 Land Drainage Act: drainage works could be undertaken if they were to be of public benefit. Despite the many problems that had beset drainage in Manitoba, he thought that the entire province had derived significant benefit. While noting that approximately 2.1 million acres were included in drainage districts and that they were among the most productive areas of the province, Lyons argued that the benefits of drainage were still far more widespread.[80] Among them, he highlighted increased land values, diversified and increased agricultural production, and enhanced transportation and communication networks. Even those Manitobans most dissatisfied with specific drainage projects recognized that, in general, a drained landscape was more productive than an undrained landscape. Complainants to the minister of public works occasionally accompanied their demands with acknowledgments that "successful farm communities" in many areas of the province "could never have been settled were it not for the drainage work."[81] Paying for drainage remained a challenge, but drainage itself had important positive consequences for the province at large. Whether an undertaking made necessary because of the particular geography of the province or an undertaking typical of the liberal tradition of supporting private enterprise, by the 1940s many agreed that drainage served the public interest.

In the early 1920s, highlanders had objected to paying a levy that would fund the construction of drains through the lowlands. Insofar as liability for the drainage tax hinged on highland land use changes, highlanders saw in the tax a violation of their right to manage their lands as they saw fit. Instead of a curtailed vision of the rights of individual landowners, Lyons's recommendations turned on an expanded sense of the responsibilities of Manitoba. Lowlanders would continue to fund the drainage works necessary to support agricultural production on their lands, but their costs would be much reduced as the province would assume the expense of foreign water drainage. Highlanders would not be obliged to make any direct financial contribution to solving a problem for which they continued to deny responsibility.

Significantly, the perceived injustice of spending provincial money on projects that targeted a particular area was part of the original impetus for developing the system of drainage districts as a means of trying to ensure that those who benefited from drainage would pay for it. In the 1890s, government expenditure on drainage was a problem that drainage districts were intended to solve; in the 1940s, government expenditure on drainage was to solve the foreign water problem that drainage districts (or, more particularly, their constrained boundaries) had helped to create. Ultimately, Lyons's proposed resolution of the highlander/lowlander dispute derived not from a more sophisticated understanding of the Manitoba environment or a more concerted effort to align land use to land conditions but from a major shift in how drainage was to be funded.

In the mid-twentieth century, funding drainage by the province at large was more acceptable than spreading the costs over the affected watershed. Although by then certainly less controversial than in the 1920s, the watershed idea had not yet become sufficiently accepted to make it possible to coordinate the funding of surface water management at that scale. There was greater political willingness to increase provincial funding of drainage than to confront the challenges inherent in tailoring land management to environmental conditions. Drainage funding that involved all in the province was acceptable; drainage programs that operated by watershed were an outrage. The watershed was still not a legitimate management unit.

The settlement boom across the Canadian Prairies in the late nineteenth century and early twentieth century was a time of profound regional transformation. Nested within this larger tale is Manitoba's own story of change, one that involves efforts to remake the landscape of the province through drainage. Examining contention in the drained landscape reveals how surface water management was a question of values. In deciding how to manage the ecological commons of surface water, Manitobans were debating how to define agricultural prosperity in the province. Was it best if individual farmers sought to maximize their own yields, no matter how their practices affected others? Or was it better to coordinate land management at a higher scale, limiting what farmers could do on their own lands in the hope of making it possible for others to succeed? Was the search for prosperity an individual or a collective undertaking? Ultimately, these were not the sorts of questions that engineers were trained to answer. Manitobans engaged in long-term debate over these matters, often without arriving at anything like resolution. Yet the process itself deserves acknowledgment. This was a wide-ranging debate, one that engaged principles basic to life on the Canadian Prairies. This was a probing examination, if a colloquial one, of land management practices in tune with liberal ideals.

## CONCLUSION

Drainage was both a concrete problem requiring a practical solution and a catalyst for critical thinking about the significance of professional expertise, the responsibilities of Manitobans to each other, and the role of the state. The expansive character of drainage, involving questions seemingly far removed from the problem of excess surface water, did not make the matter any easier to resolve. In some ways, the passage of the 1895 Land Drainage Act transformed drainage in the province. It established the necessary financial and administrative structures to support larger projects more consistent with both the scale of the drainage problem and the private property ideal prevalent on the Prairies. In other ways, the legislation became a fundamental aspect of the drainage problem, for it provided the means to create drainage districts without regard for the watershed boundaries of the province.

The Land Drainage Act mandated involvement by the province's environmental professionals, signalling the growing importance of the contributions of engineers and surveyors. The government's response to two decades of unsuccessful drainage represented real change in provincial administration as the patronage practices of earlier days were replaced by new expert-led decision-making processes. In light of the urban focus of some scholars concerned with the growing significance of professional expertise, it is notable that in Manitoba this transition took place partly in relation to change in the rural context. To invoke a concept elucidated by environmental historian James Murton in his analysis of BC land settlement programs, this was a countryside further modernized through engagement between farmers and experts.[82]

In the late 1910s and early 1920s, government-commissioned experts provided an explanation for continued flooding and a solution in watershed-based drainage funding. But there was no broad legitimization of the watershed idea. Indeed, the proposed solution catalyzed the creation of a highland community of interest, defined largely in opposition to the lowland perspective that had been generated years earlier through shared vulnerability to flooding and investment in potential solutions. Even as lowland Manitobans embraced the idea of watershed-based drainage funding as a means of securing the funds necessary to address persistent flooding, the idea found little purchase in a highland landscape of progress and property. Out of highlander/lowlander debate before the Legislative Committee on Drainage emerged the idea that the entire province should fund drainage. There seems to have been no direct link between this creative moment and the parallel recommendation of Lyons nearly thirty years later, but the moment itself is still noteworthy. Prompted by challenging environmental conditions, Manitobans considered environmental management arrangements that would in fact be adopted in decades to come. It was ultimately more viable to shift the financial burden for draining highland water to the province than to enact a contentious arrangement such as watershed-based drainage funding.

Before the Legislative Committee on Drainage in the early 1920s, J.G. Sullivan asserted that drainage was "one of the most complicated and difficult problems in the country or in the world since the beginning of time."[83] Although melodramatic, such an assertion might have provided some comfort to Manitobans who found themselves deeply mired in debates over drainage. In grappling with the geography of their province, Manitobans confronted not only the environmental conditions but also the political and social arrangements that bore on their lives. As they debated drainage, they considered the character of their society. In exchanges that took off from drainage, Manitobans grappled with broad questions that were posed in particular ways by their natural and cultural environments.

# 4

# International Bioregions and Local Momentum: The International Joint Commission, Ducks Unlimited, and Continued Drainage

From the earliest days of resettlement through the 1930s, most Manitobans favoured drainage. The agricultural development of southern Manitoba depended on the production of dry land. Although the specifics of drainage were highly contentious, with vigorous debate turning on the hows and whys of particular government projects, the idea of drainage itself was hardly controversial. In the wet prairie, drainage was the means to bring the local landscape in line with the agricultural ideal, to extend crop production across the patchy wetlands of the region, and to ensure that intermittent flooding did not interfere with the growing season. Drainage amounted to the efficient use of an inefficiently laid-out landscape. It was the government-assisted improvement that, though satisfactory to nearly no one, was desired by nearly everyone.

A few decades into the twentieth century in Manitoba, the consensus in favour of drainage began to erode. The mindset that emphasized efficiency in the use of natural resources such as land continued to prevail. But in the wet prairie as elsewhere, the specific understanding of this concept began to shift.[1] By this time, some people involved in land management in Manitoba began to look more closely at the meaning of efficiency in relation to the local landscape. How should drainage be managed? Was drainage really the best way to make use of all parts of the wet prairie? Answers to these questions are suggested by the histories of water management on the Roseau River and in Big Grass Marsh.

The Roseau River crosses from the United States into southeastern Manitoba before joining the Red River some 22.5 kilometres (14 miles) into Canadian territory. This region was flooded periodically, but the situation deteriorated shortly after the United States and Great Britain (on behalf of Canada) signed the Boundary Waters Treaty in 1909. The border between Canada and the United States runs approximately 8,900 kilometres, of which some 3,500 kilometres traverse water bodies.[2] Resulting from recognition of the need to prevent conflict over transboundary waters, the Boundary Waters Treaty addressed issues such as flow diversion and pollution. It also created the International Joint Commission (IJC) as a mechanism to facilitate coordinated management of boundary waters.[3] Because of the international aspect, the dominion government became far more involved in surface water management along the Roseau River than in other parts of the province. The establishment of the IJC and its hearings on the management of the Roseau River expanded the number of actors involved in drainage and resulted in a significantly different approach to a water management problem not unlike many still unresolved water management problems in the wet prairie.

Big Grass Marsh, situated west of the southwest shore of Lake Manitoba, was once among the largest wetlands in Manitoba. But here, as in other places, many newcomers saw not a productive wetland ecosystem but a potential agricultural landscape. By the 1930s, after intensive drainage failed to produce valuable agricultural land, the area was more a moonscape than a marsh. Its peat soils were highly combustible, and fires smouldered underground. Escaping smoke mixed with the soil that drifted across the landscape, creating an atmosphere of eerie neglect. But to some ambitious Americans and their Canadian collaborators, this seemed to be the ideal location in which to launch their newest enterprise. In the late 1930s, Ducks Unlimited (DU), an international conservation organization with American roots, brought water back to the waste land of Big Grass Marsh to create Duck Factory No. 1. A short biography of the marsh reveals the organization's restoration efforts and the intellectual consequences they had. Ultimately, efforts to bolster duck populations contributed to changes in public perceptions of prairie wetlands.

Manitoba involvement along the Roseau River and in Big Grass Marsh connected the province to interests that spanned geopolitical borders. Along the Roseau River, floodwaters involved both Canadians and Americans in negotiations over water management; to the west of Lake Manitoba, international waterfowl migration patterns meant that foreign waterfowl enthusiasts had an interest in prairie wetlands. The presence of these bioregions, geographical entities defined by a particularly significant ecological element, linked Manitoba to other places and proved to be important to change in surface water management. Ultimately, newly interested parties from outside Manitoba introduced fresh ways of thinking about surface water management, affected local environmental thought, and suggested new ways of defining efficiency in the local environment.

But new ways of perceiving the Manitoba environment did not necessarily amount to changes in provincial drainage practices. Even as government officials joined Ducks Unlimited, lending their public personas to campaigns for wetland restoration, significant government investment in drainage continued. No one paused to think that draining a farm field while restoring a neighbouring wetland might be perverse. What had once been a broad consensus over drainage was beginning to break down, but institutional inertia limited the extent to which changed perceptions of wetlands affected provincial government drainage practices. Bioregional links prompted some instances of innovative environmental management, but the local political context proved to be resistant to change. Water flowed through the Manitoba landscape, but new ideas did not diffuse so readily.

## ROSEAU RIVER AND THE INTERNATIONAL JOINT COMMISSION

If North America could be abstracted from its political and economic history, there would be as many reasons to divide it into long north-south strips as into the east-west blocks of Canada and the United States. Indeed, a number of commentators have remarked on the north-south grain of the American continent.[4] Political scientist Kim Richard Nossal has suggested that this orientation is reflected in the forty-two rivers that cross and crisscross the 8,900-kilometre (5,500-mile) border between the two countries.[5] In Manitoba, the most renowned of these rivers is the Red River. Yet the much smaller Roseau River figured large in early international cooperation on water management. The Roseau runs northwestward through the state of Minnesota for about 65 kilometres (40 miles) before crossing into Manitoba. It drains an area of approximately 5,335 square kilometres (2,060 square miles), slightly less than half of which are in Canada.[6] In 1893, Chief Engineer Louis Coste with the dominion Department of Public Works reported that the river was 30 to 60 metres (100 to 200 feet) wide and varied in depth from 0.5 metres (1.5 feet) in dry times to 3 metres (10 feet) in wet periods. He also noted that in some places the river was particularly marshy, with vegetation growing as high as 2 metres (7 feet).[7]

The lands along the Roseau River's Manitoba reach were not among the province's most desirable for agriculture.[8] The soil was stony and fairly unproductive. Although a scattering of settlement was established there over Manitoba's first decades, the land was, according to an 1896 report of the Department of the Interior, "not such as to attract much attention from the Canadian settler, being to a good extent very rough and hard to clear and improve."[9] But the area came to the attention of some who saw things differently. The local combination of abundant wood and water, along with plenty of room for community expansion, proved attractive to newcomers from Central and Eastern Europe. Pockets of ethnic concentration developed, including a significant number of immigrants from Ukraine.[10]

Over time, mixed farming communities grew up. A government official noted that, for the "hard-working, thrifty type of Central European" who had settled in the area, "the revenue from stock, dairy products and hay provided a good living for a family on nearly every quarter section."[11] Extensive grain farming was not the goal of these settlers; rather, they preferred to replicate the intensive cultivation of small areas that had taken place in their homelands. As historical geographer John Lehr explains, even those "who selected a quarter which was largely swamp still felt confident that twenty acres or so of good land was more than adequate for their needs."[12] The newcomers pursued their vision of the good life, and community institutions continued to develop. Two municipalities were formed along the Roseau River: Franklin in 1883 and Stuartburn in 1902.

Prior to extensive Ukrainian settlement, the provincial government had already recognized that farms in the area would be threatened by insufficient drainage and periodic flooding. Improvement projects along the Roseau River began in 1881. A drain was built to the river about 1.5 kilometres (1 mile) north of the international boundary. Due to both faulty design and poor execution, little land was improved.[13] Sent to evaluate the situation, engineer John Molloy warned that failure to alleviate flooding would likely result in abandonment of the region.[14] Even farmers satisfied with twenty usable acres could find themselves in trouble as flood patterns shifted and small patches of cultivation were threatened. The Rural Municipality of Franklin undertook numerous drainage projects, accumulating a debt of more than $100,000.[15] This initiative proved insufficient. Officials with both the province and the municipality came to believe that formation of a drainage district under The Land Drainage Act was the best way to secure financing for the necessary works. In 1899, Drainage District No. 3 was formed.

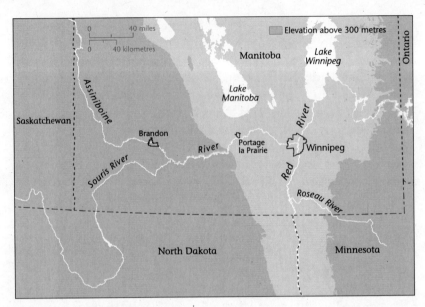

FIGURE 16   Roseau and Souris Rivers

Even generous funding and careful construction failed to solve the drainage problems of the area. Some Canadians insisted that flooding was due to changes in the watershed south of the forty-ninth parallel. But there, as elsewhere, there was no consensus on ecological linkages between the upper and lower watersheds. In the United States, what has become known as the forest-stream-flow controversy, a debate over flood control strategies that raged with particular fervour from the Progressive period through the New Deal era, turned partly on whether and to what extent deforestation altered runoff patterns. Could floods be mitigated adequately through changed land management and small dams? Or was it necessary to construct massive water control structures? American federal agencies offered conflicting opinions about how action in the upper watershed affected conditions down below. The Department of Agriculture advocated careful land management as an effective flood mitigation strategy, whereas the Army Corps of Engineers argued that only dam construction offered real security. Ultimately, each lobbied for a conception of the problem of flooding that corresponded to the solution it offered.[16]

As federal money and agency prestige hung in the balance, the forest-stream-flow controversy was fought with fervour comparable to that evident in the disputes between highlanders and lowlanders in Manitoba.

Although the issue in the province was the downstream consequences of agriculture rather than deforestation, in both cases uncertainty proved to be conducive to intense and persistent conflict. Flood-vulnerable Manitobans argued that highland contributions to drainage funding were only fair given how land use changes up above exacerbated flooding down below. Highlanders denied any such liability. By the 1920s, these divergent perspectives had hardened into fundamentally different ways of perceiving the landscape. This makes it all the more remarkable that, when the upstream/downstream question was recast across the international border as a conflict between Minnesotans and Manitobans, there emerged a significantly different way of managing the situation.

As the 1909 Boundary Waters Treaty inaugurated a new era of international cooperation on waterways of mutual concern, American officials suggested to Manitobans that both parties would benefit through cooperative management of the Roseau River.[17] The Roseau also offered an opportunity to address contentious issues at a relatively small scale.[18] If a satisfactory resolution to transboundary management problems could be found, the precedent might govern the resolution of similar issues along larger waterways, such as the Red River.[19] Despite continued problems along the Roseau and the opportunity to establish a precedent, American overtures for cooperation on Manitoba's boundary were largely ignored until the 1919 creation of the Sullivan Commission. Only then did Manitoba officials begin to reflect seriously on the state of the drained landscape, the sort of stock taking that was preliminary to any consideration of what might be gained through an attempt to coordinate environmental management across the international border. Only at that point did the international bioregion created by surface water begin to affect surface water administration in Manitoba.

In December 1919, American representatives travelled to Winnipeg to meet with politicians and administrators from the governments of Manitoba and Canada. The idea was "to arrive at some basis of co-operation on the large international aspects of the Red River problem."[20] In attendance were J.G. Sullivan and J.A. Thompson of the Sullivan Commission and H.A. Bowman, chief engineer of the Manitoba Department of Public Works. This meeting did not, however, result in any concrete water management strategies, and no ongoing relationship was established between Canadian and American officials. The Sullivan Commission had been created to examine a domestic problem. Its influence on drainage policy was limited, even on questions that fell squarely under its purview. On the international matter, as with so many others, the progressive activity of the Sullivan Commission had little real impact.

This was partly attributable to ongoing jurisdictional conflict. Both the province and the dominion were adamant that the other had lead responsibility. The province argued that the international dimensions of the Roseau River issue made it a dominion concern. Ottawa countered that the matter was provincial jurisdiction under the British North America Act because the Roseau was non-navigable and the problem was fundamentally one of drainage. Dominion officials might also have been deliberately uncooperative to protest Manitoba's continued lack of interest in a drainage arrangement with Ottawa comparable to that negotiated with Alberta and Saskatchewan.[21] There is also evidence that dominion officials were concerned about their authority in the region of southern Manitoba that was excluded from the 1894 Northwest Irrigation Act.[22] The difficulty of sorting out which government should take the initiative was compounded by recognition that the Americans likely would contest any claim that upstream drainage exacerbated downstream flooding. Facing jurisdictional ambiguity and anticipating international disagreement, Canadian and Manitoba officials were disinclined to establish an ongoing dialogue with their American counterparts.[23]

While Canadian governments resisted American overtures, drainage in the northern states continued. Most drainage in Minnesota took place in the first decade of the twentieth century, altering the conditions that Drainage District No. 3 had been formed to address.[24] By 1920, there were 90,446 drained farm acres (36,602 hectares) in the American portion of the Roseau River watershed.[25] Some Canadians concluded that drainage by Americans had altered the magnitude and timing of flow in a manner detrimental to people living along the river north of the border. In 1919, J.M. Mysyk, secretary-treasurer of the Rural Municipality of Stuartburn, informed the Department of the Interior that "the [Minnesota] government is dredging the swampy countries and lets all the water flow into the Roseau River which is too small in its bed to carry that mass of water." Flooding resulted. "The settlers living in the neighborhood," he elaborated, "have completely lost their coming crops and a cry is heard from them everywhere."[26]

There was no immediate response to Mysyk's protests. In 1925, ninety-nine settlers in the Roseau River area expressed a desire to abandon their lands and take up new homesteads elsewhere.[27] This got the government's attention, and a dominion agent was dispatched to collect additional information on the environmental reversal caused by exacerbated flooding.

Under the influence of excessive water, "a change began to take place." The agent explained how

> The land began to go back to its former barren state. The hay sloughs, usually passable by July first, remained wet all summer, and in a few years once profitable stands of grass began to be replaced by a wire grass and rush growth. Wire grass areas began to be replaced by black rush and moss, while the shallow sod began to disappear and the sand and gravel to appear on the surface.[28]

Government officials sent to evaluate the situation concurred with local residents: the community's agricultural progress was at risk of being washed away because of changes in flood patterns, and these patterns had changed because of American drainage. With the claims of locals confirmed, governments were moved to cooperate. Manitoba and Ottawa agreed to share the cost of constructing a substantial double dyke floodway designed to alleviate flooding north of the international border.[29]

Manitoba and the dominion were collaborating, but they were still not prepared to work with the Americans. Canadian reluctance to engage in transboundary negotiations is reflected in the design of the double dyke floodway. Plans were deliberately modified to eliminate any possibility of a backwater effect that would alter conditions on the American side of the border.[30] The hope was that careful engineering could fashion a domestic solution to an international problem, that the bioregion could be severed in accord with geopolitical boundaries. Despite the best efforts of Canadian engineers, American experts doubted the project would leave the upper watershed unaffected. A reference to the IJC was again proposed, and this time Canada chose to cooperate.

As the dominion government took the lead on the Roseau River matter, Winnipeg began to assert its interest. Urban development near the forks of the Red and Assiniboine Rivers depended on drainage, but urban drainage differed from rural land drainage.[31] The city was not a player in most early debates over agricultural land management. But the severity and timing of high-water episodes along major waterways that ran through the city such as the Red and the Assiniboine could be affected by changed drainage patterns in the agricultural landscape. Civic administrators recognized that, as rural land drainage might bear on urban flood patterns, Winnipeg had an interest in conditions upstream. Perhaps prompted by J.G. Sullivan's 1922 article in the *Engineering News-Record* suggesting that drainage in the lower Red River valley would likely worsen conditions in the city, Winnipeg officials of the mid-1920s investigated urban flood protection.[32] In 1927, in light of attention to flooding in the provincial south, City Solicitor J. Prudhomme approached the dominion government. He was concerned that projects might have a deleterious effect on the city.[33]

In a bid to serve Winnipeg, Prudhomme argued before the International Joint Commission that the city had a right to compensation for any damage caused by flooding of the Red River because of changes to the Roseau River watershed. His brief made clear that "communities which desire to change the natural conditions must not be allowed to do so at the expense of other communities without proper steps being taken" to provide appropriate protection or compensation.[34] Winnipeg might be affected as the Roseau conveyed excess water to the Red and the Red conveyed it to the city. Prudhomme's position paralleled the arguments put forth by lowlanders who argued before the Legislative Committee on Drainage that they should not suffer the consequences of highland changes that modified the flow regime. Winnipeg was a new actor in the drainage debate in the late 1920s, but its representative invoked an old logic.

For those participants in the IJC hearings more attuned to the international context, Winnipeg's argument was not useful. Within Manitoba, water always flowed from highlander farms to lowlander farms; between Canada and the United States, rivers ran both ways. As Kim Richard Nossal has explained, rather than a consistent upstream/downstream relationship between Canada and the United States, "there are upstream and downstream *localities* in both states along the length of the frontier."[35] There was no standard direction of flow across the Canadian-American border; neither one was the upstream or the downstream nation. Although downstream along the Roseau River, Canadian officials were careful not to invoke arguments that could be turned against them in other contexts. They were particularly fearful of compromising their position with respect to the Souris River.

The Souris River runs southward from Saskatchewan into North Dakota before it curls northward into Manitoba to join the Assiniboine River. The land along the river's 700 kilometres (435 miles) is rich, flat farm country. Although the Souris passes through drier land than the Roseau, wet periods could still lead to flooding. In April 1928, Americans requested information about drainage in Saskatchewan on the ground that drainage along the Souris River was worsening flooding in Minot, North Dakota. In response, J.T. Johnson, director of the Water Powers Branch of the Canadian Department of the Interior, supplied the requested documents and indicated that he would like to receive comparable information about American drainage in the Roseau River area.[36] The tone of these letters was entirely diplomatic, but the subtext was clear. Any action that Americans took with regard to flooding along the Souris would be matched by Canadian action along the Roseau.

Although the success and longevity of the IJC has been linked to how commissioners investigated each reference independently,[37] officials in each country made connections between localities in ways that served their national interests. Thus, Canada's representatives sought a compromise that seemed to be fair in light of the current condition of the Roseau River and the needs of the communities situated along its course. For those with an eye to conditions in the different localities situated along the international boundary, it was important to balance the effects of alterations in one place with the effects of alterations in another.[38] Their approach was more conciliatory than that of Winnipeg, for it was moderated by recognition of the challenges of defining fairness in dynamic natural and cultural contexts. This approach provided neither a straightforward means of dispute resolution nor any obvious answers to the difficult question of who exactly should contribute to the costs of surface water management. But participants were left to grapple with the environment before them in a context more conducive to open dialogue and careful compromise. The polarization of views that took place between highlanders and lowlanders in Manitoba (or between federal agencies in the United States, with respect to the forest-stream-flow controversy) was not replicated across the international border.

Ecological linkages between upper and lower watersheds remained poorly understood. This was evident at the IJC hearings into the Roseau River matter, particularly in an exchange between American official E.V. Willard and IJC Commissioner Charles A. Magrath. Willard explained that "the popular notion is that drainage ditches accelerate the runoff and in that way increase the peaks of floods." Magrath pressed for greater certainty, asking whether this was a fact. Willard prevaricated, claiming he "would not say so as an unqualified statement."[39] As an American drainage expert, Willard likely realized it served his national interest to qualify the general principle. But even decades later, in the mid-1970s, Canadian government officials noted that the ecological linkages between highlands and lowlands remained under dispute.[40] And beyond this issue, there was also a real difficulty in reconciling general principles and specific locations. Any sort of financial compensation would require a precise understanding of cause and effect in a particular place, including a measurement of increased surface water flow. This would not be easy to achieve. What is remarkable about the Roseau River situation is how uncertainty served to prompt compromise rather than polarize debate. The management of international waterways suggested another framework for discussion about how to live within a watershed, one that emphasized the need for reasonable solutions that were fair, as far as possible, to all parties. In Manitoba, perceived human-caused environmental change was the spur to intense debate between highlanders and lowlanders over whether change occurred and whether financial compensation was due. In the international context, perceived human-caused environmental change was the starting point for negotiations.

Interestingly, Sullivan foresaw the parallel between highlander/lowlander disputes and those between Americans and Canadians. In trying to convince highlanders to accept that they should contribute to the cost of lowlander drainage, Sullivan invoked the American example. If upstream Canadians failed to accept liability, he warned, then upstream Americans might build on this precedent to avoid paying compensation to Canadians detrimentally affected by American drainage.[41] There was some political attention to this argument, and the recommendations of the Sullivan Commission were defended by the Manitoba public works minister W.R. Clubb in light of what he asserted at this stage was the obligation of Minnesota to compensate Manitoba for damage caused by American drainage.[42] Clubb was arguing for consistency of policy on domestic and international water management questions, and, unless Manitoba was willing to overlook damage to provincial lands by American drainage, this meant holding highlanders responsible for some of the costs of lowland drainage. However, as Manitoba kept out of the international arena, most provincial politicians were able to avoid grappling at this point with the implications of the parallel between drainage across the international border and drainage within the province.

In a report released on 8 June 1929, the IJC ruled that works under way along the Roseau River north of the international border would not worsen conditions in the United States.[43] With this matter resolved, Canadian flood mitigation projects continued. The situation along the Roseau was further eased by the dry years of the 1930s. District Engineer F.G. Goodspeed included a vivid description in his August 1929 letter to Chief Engineer K.M. Cameron:

Everything is very dry, the river itself carrying scarcely any water. We were able to walk across the river on some stepping stones which had been placed for that purpose. We also motored almost the same ground that Mr.

Corriveau and I crossed in a canoe a year ago. The ground is absolutely dry and parched and the grass crunches under foot although appearing green on top.[44]

Completed diplomatic processes and changed environmental conditions meant that government officials could turn their attention elsewhere. But if this dry landscape had Canadians along the Roseau River breathing a sigh of relief, over subsequent years few others would find much to celebrate in similar conditions. Canadians and Americans who lived along the Souris, which ran through drier country than the Roseau, became more concerned with water scarcity than flooding. Even the flow of the Red River was substantially reduced in this period.[45] The drought of the 1930s was an environmental disaster across much of the Canadian Prairies. Its consequences for water management in Manitoba were significant.

## Ducks Unlimited and Manitoba's Big Grass Marsh

Warren Upham, a geologist with the US Geological Survey and the author of an 1895 volume on the glacial Lake Agassiz region, described Big Grass Marsh prior to extensive drainage. It extended parallel to Lake Manitoba's western shore for "more than 20 miles [32 kilometres] from south to north, with a width of 3 to 5 miles [5 to 8 kilometres]." Both the White Mud and Big Grass Rivers ran into the area and then "flow[ed] sluggishly through a broad, quaking morass" interspersed with "shallow, rush-filled lakes."[46] From the late nineteenth century to the early twentieth century, aspiring drainers – settlers, capitalists, or government officials – saw potential in the region. Later on, when it was clear that no amount of ambition would translate into successful reclamation, the restoration of Big Grass Marsh by Ducks Unlimited figured in the development of the conservation idea in Manitoba. As DU agents matched their physical labour in the marsh with publicity efforts in the public sphere, they promoted a view of wet areas, whether in Manitoba or elsewhere, as more than submerged farmland.

For newcomers, Big Grass Marsh inspired big dreams of environmental transformation. In 1873, settler William Gordon, writing on behalf of several homesteaders in the region, sought dominion assistance to improve water flow through the marsh. This petition urged the government to support settlers' drainage efforts, both to aid those already there and to attract more immigrants. The petitioners were optimistic that the project could be accomplished at a reasonable price. About $100 would take care of the most pressing work; for $300 or $400, the entire swamp could be drained and made available to settlers.[47] It is not clear if the dominion government was convinced by these arguments or if the local people went ahead with their project. Abundantly evident, though, is that the drainage of Big Grass Marsh proved to be a much more complicated undertaking than it seemed to these optimistic early settlers.

Although early settler appeals had been directed at the dominion, the provincial government soon became involved in the drainage of Big Grass Marsh. The 1880 drainage lands arrangement between Winnipeg and Ottawa was designed to remake just this sort of substantial marsh.[48] Under its terms, title to most of the even-numbered sections in areas improved by the province would be transferred from Canada to Manitoba. Big Grass Marsh was among those areas the province proposed to drain.[49] In May 1880, engineer G.B. Bemister was sent to determine what would be necessary to drain the marsh.[50] He reported that the work would be challenging because of the size of the project and the difficulties of working in a swampy location. Not easily dissuaded, the provincial government contracted a number of private parties to excavate ditches through the area.[51] Under the drainage lands arrangement, funds for the work would come from Manitoba's capital account with the dominion government.[52] The province would then lend the money to private companies that would undertake the drainage, on the understanding that the loan would be repaid in full six months later.[53] The Manitoba Drainage Company won the contract for Big Grass Marsh.[54] The company divided the undertaking into smaller projects and subcontracted them to local labourers at low rates, retaining as profit the difference between the amount received from the government and the cost of the subcontracts. Additionally, a portion of the area the dominion transferred to the province passed directly to the company.

The Westbourne Municipal Council was happy to have the Manitoba Drainage Company take charge of local drainage, as long as "the main object" of eliminating flooding was achieved.[55] However, it eventually became clear that the ditches dug had not been sufficient. Although years passed without additional large-scale drainage undertakings, complaints from flooded residents kept the matter alive. Seizing the new opportunities offered by the 1895 Land Drainage Act, the municipality lobbied for organization of a drainage district in Big Grass Marsh.[56] Flooded residents were disheartened, however, by the results of a provincial survey that concluded the area would be extremely difficult to drain thoroughly.[57] Bounded on the east by a pronounced ridge that ran north-south about four or five miles from the lake, Big Grass Marsh presented particular

FIGURE 17    The excavation of a drain in Drainage District No. 8. The scrubby vegetation
evident in undisturbed areas was typical of Manitoba's parkland region. Drainage District
No. 8 included Big Grass Marsh.
*Source:* Archives of Manitoba, Drainage 27, 1911, A.R. Boivins Drainage 60 D. D. No. 8.

challenges to drainers.[58] As neither cutting through the ridge nor carrying
the water around it seemed to be feasible, "the scheme of forming a drainage district was abandoned," at least temporarily.[59]

In 1909, perhaps as a result of increased demand for agricultural land,
caution was set aside, and Drainage District No. 8 was formed in Big Grass
Marsh.[60] Work started in August of that year.[61] The project included the
deepest channel ever cut for drainage in the province; in places, it was four
metres (thirteen feet) deep.[62] As the years passed and the work continued,
the government received both protests against the drainage and requests
that it be expanded. The project was enlarged in 1912. More than $880,000
was spent in Drainage District No. 8.[63] A local paper proclaimed that "the
province has gained thousands of acres of land which will produce returns
on a valuation of millions of dollars." The loss of "a grand duck shooting
ground at Big Grassy Marsh" was worthy of mention but certainly was not

a reason for second thoughts about the wisdom of drainage.[64] The anticipated agricultural value of the reclaimed lands trumped whatever leisure
or nutritional value had previously been derived from bagged waterfowl.

But the local press was too quick to count up gains and losses. Not all
drained land in Big Grass Marsh equated to agricultural land. Some lands
around the marsh edges were successfully cultivated, but about 35,000
acres in what had been the lowest part of Big Grass remained unsuitable
for crops.[65] Peaty soils were the major problem. In 1934, the Committee
on the Utilization of Public Lands in the Province of Manitoba argued
that the area should never have been brought under The Land Drainage
Act.[66] The following year a land survey sponsored by the provincial government indicated that only 4 percent of the lands in Drainage District No.
8 were likely to support successful farms focused on crops. Slightly less
than a third of the area was considered adequate for farms relying on
livestock as well as crops. The rest of the land, over 260,000 acres (105,200
hectares), was entirely unsuitable for agriculture.[67] In contrast to the expectations of settler William Gordon, who had anticipated that drainage
would attract agricultural settlers, the region was losing population at a
rate that distinguished it from other parts of the province.[68]

Although drainage had not produced farms, water patterns had been
altered dramatically. With the water table lowered, formerly dependable
wells began to run dry.[69] For the first time, residents were obliged to haul
water. Marsh grasses valuable to cattle ranchers and sheep herders no longer
grew. Although the area remained vulnerable to flooding in especially wet
years, in most years it was little more than barren land. Peat fires and dust
storms were particularly severe in dry periods. At points in the 1930s, there
were more than three hundred fires burning in the former marsh.[70] In
these droughty years, some drainage canals deteriorated and filled with
sand and ashes.[71] Few owners were able to make profitable use of their
land, and much of it reverted to the government.[72] In the late nineteenth
century, governments referred to wetlands as waste lands. Drainage was
perceived as a way of creating useful lands. In a striking reversal, in 1938
the author of a letter to the editor of a local newspaper described the
drained marsh as "a burning waste."[73]

In the late 1930s, this singularly unsuccessful drainage district seemed
to typify the state of much agricultural land across the Canadian Prairies.
Although Manitoba had little to compare with the drought-stricken areas
of south-central and southwestern Saskatchewan and southeastern Alberta,
southwestern Manitoba was of a piece with southeastern Saskatchewan,
and the situation west of the Red River seemed to be comparable to that

of northern Saskatchewan.[74] Such widespread hardship was conducive to what Manitoba's Department of Mines and Natural Resources described in 1937 as a rise in conservation sentiment.[75] Although the fate of farmers remained paramount, some people also began to consider another group that had been hit hard by the drought: ducks. If Big Grass Marsh seemed to be a picture of agricultural failure, it was also "a death trap for waterfowl."[76]

For those with a long acquaintance with the area, the dearth of waterfowl was as shocking as it was troubling. Spring and autumn goose hunts had been fixtures in the Red River settlement. As described by settler J.J. Hargrave,

> Many families leave the settlement and go off a distance of sixty or eighty miles to the neighbouring lakes to live for a few weeks a camp life in the open air. The geese which fly with almost incredible speed and at great height come down to drink from the lakes and rivers, on the shore of which the hunters are encamped, and are dispatched by the latter in great numbers.[77]

The seasonal pattern of exploitation persisted through the nineteenth century as settlers matched their hunting forays with the seasonal movement of birds along what people today call the Mississippi flyway.[78] It was the wetlands that drew in the geese that flew so high and so fast; it was also the wetlands that provided food and shelter to the migrating geese. Wild game remained a key source of dietary protein through World War I, supporting a market for waterfowl that, as in other parts of Canada and the United States, affected duck and geese populations even as it kept people fed.[79] By the early decades of the twentieth century, waterfowl numbers were in serious decline.

In the United States, concern about waterfowl populations had been mounting for decades. Overhunting and habitat loss were identified as major factors, and international agreements such as the Migratory Bird Treaty, signed by the United States and Great Britain (on behalf of Canada) in 1916, reflected early efforts to limit depredations and protect breeding grounds.[80] But waterfowl populations remained a concern, and the Manitoba government was led to consider ways of ameliorating the situation.[81] Although Big Grass Marsh epitomized the agricultural problems evident across the Prairies, it also held particular potential within the continental project of waterfowl rehabilitation. Yet though provincial officials thought by the mid-1930s that their province was "by far the most important

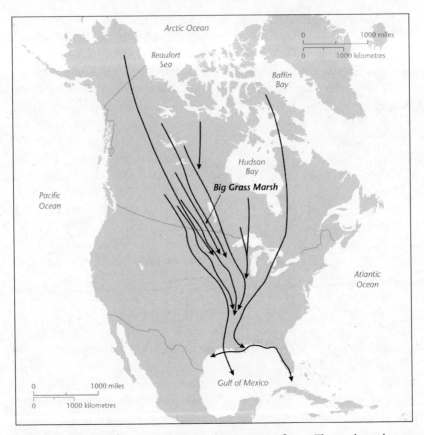

FIGURE 18    Big Grass Marsh in relation to the Mississippi flyway. The north-south lines represent the major migration routes that together make up the flyway.
*Source:* Adapted from W. G. Leitch, *Ducks and Men: Forty Years of Co-operation in Conservation* (Winnipeg: Ducks Unlimited Canada, 1978), 76.

breeding ground for game wildfowl," they were also aware that "the problem of safeguarding wildfowl is not a local one."[82] There was a significant administrative challenge derived from the fact that "wildfowl are cosmopolitan creatures and their welfare is in the interest of all provinces and states in which it seeks seasonal refuge."[83] Conservation efforts manifested through state agencies were complicated since waterfowl migration patterns spanned geopolitical boundaries. The Mississippi flyway, for instance, was a bioregion that extended across the border between Canada and the United States. Those interested in migratory birds along this middle reach of the continent were interested, whether they realized it or not, in the condition of Big Grass Marsh.

In October 1930, a group of hunters in the United States formed the More Game Birds in America Foundation. Their intention was to apply the same methods to duck propagation as they believed had led to their success in the business world. The foundation harnessed burgeoning enthusiasm for wildfowl conservation and quickly made its influence felt on the American scene by establishing connections with politicians, scientists, and conservationists.[84] However, just as the Manitoba government thought the success of any conservation efforts would be limited by provincial borders, so too the More Game Birds in America Foundation ran up against the forty-ninth parallel. Convinced that waterfowl numbers depended on the preservation of habitat in Canada, American activists began to develop a method by which they could work for the conservation of Canadian wetlands.

The first step was establishing connections with similarly minded Canadians, including lawyer and later judge W.G. Ross, newspaper executive O. Leigh Spencer, and entrepreneur James Richardson.[85] The second was a large-scale census of the duck population, conducted in the early 1930s through cooperation between Canadians and Americans.[86] By 1935, conservationists thought they had confirmed that "the future of waterfowl lay in preserving the breeding grounds in Canada."[87] They devised a plan involving the preservation of unspoiled habitat and the restoration of degraded areas and created a new organization to bring this about.[88] Ducks Unlimited was incorporated on 29 January 1937; Ducks Unlimited (Canada) followed on 10 March 1937. Thomas Main, the newly appointed general manager of the Canadian operation and a former Canadian National Railways surface water engineer, described the relationship between the two organizations: "The function of DU, Inc., is to accept cash from US sportsmen and transmit it to DU (Canada)," he explained. "The purpose of DU (Canada)," he continued, "is to invest that cash in restoration work in the Canadian West – to put bigger crops of ducks on the wing."[89] Together, the agencies would make possible a flow of cash to match the flight patterns of migratory birds.

Early on, DU had trouble raising money. In the first year, fundraising efforts fell significantly short of the target.[90] Part of the problem, at least as the agency itself saw it, was that not all potential donors "had a real conception of the breeding picture."[91] The expenditure of American money on Canadian projects was justified through an emphasis on efficiency: it was claimed that "at least five times as many ducks" would result from each dollar spent in Canada as would result if the same dollar were spent in the United States.[92] Those working for DU thought that reluctance to donate money stemmed from a lack of understanding of the underpinning ecological principles. In 1938, Harold W. Story, chair of Ducks Unlimited (Canada), gave voice to the emerging consensus – "an educational program" was needed.[93]

Whether outreach efforts were conceptualized as education or as publicity, DU was more than willing to accept their necessity. Agency officials undertook speaking tours, addressing game associations across North America and granting interviews to local newspapers along the way. Engaging printed materials were generated, some of which included direct references to Big Grass Marsh amid the anthropomorphizing and jocularity typical of many DU publications (see Figure 19). The agency also undertook a series of radio broadcasts on the topic of conservation, and examining this series reveals important aspects of the organization's outreach activity. The series was conceived as a run of fifteen-minute talks to air on a number of Canadian networks, including that owned by James Richardson, who was one of four Canadian Ducks Unlimited (Canada) directors in 1938-39.[94] The uniting theme was "Rebuilding Our Canadian West – by Conservation." The national significance of the west was emphasized, and organizers hoped that the series would build momentum that could be translated into an annual event such as National Conservation Week. Addresses were to be delivered by the premiers of the prairie provinces, other officials from the federal and provincial governments, and employees of Ducks Unlimited. All speakers were provided with an outline of approximately two pages in length that they were invited to augment with reference to their experience and expertise.

Although DU did not disguise its primary concern with waterfowl, its educational campaigns embraced a broader view of conservation. Thus, the outline – titled "Turning Manitoba Resources into Wealth" – prepared for the radio address given by J.S. McDiarmid, the Manitoba minister of natural resources, suggested that he emphasize how all resources are linked: "A marsh, developed as a wildlife production centre, will put birds in the air for miles around; yield fur crops; help replenish surface and ground waters; stabilize flowing rivers; check fires; attract tourists; and increase reservoirs for power development."[95] The prosperity of the west, and by extension of the nation, was seen to hinge on the management of natural resources from a perspective designed to ensure "the greatest good for the greatest number over the longest time."[96] In an introductory broadcast, William G. Ross, president of Ducks Unlimited (Canada), made clear why the organization agreed to do the "spade work" to put the series together. It was through conservation that its objective – "bumper crops of wild

FIGURE 19    Ducks Unlimited publicity material.
Note the references to Big Grass Marsh in the main text and along the right of the page.
*Source:* Archives of Manitoba, Edgar S. Russenholt Fonds, P 2829 15, Ducks Unlimited –
Drawings.

ducks" – could best be guaranteed.[97] From the beginning, DU believed
that fulfilling its mandate of putting bigger crops of ducks on the wing
would require a deliberate effort to explain the waterfowl life cycle.

Increasing public understanding of ecological relationships was seen as fundamental to its work. While government officials involved in the Roseau River matter grappled with the complex process of negotiating fair water management solutions in the context of environmental uncertainty, Ducks Unlimited promoted waterfowl conservation by producing entertaining publicity materials that explained what it presented as the facts about ducks and wetlands.

Ducks Unlimited worked as hard in the marsh as on the airwaves. In 1968, William G. Leitch, the organization's chief biologist, claimed that prior to its involvement Big Grass Marsh had been "an ideal example of the misuse of land and water."[98] In the 1930s, even advocates of improved drainage throughout Manitoba's wet prairie recognized that Big Grass Marsh required a different strategy.[99] Given the extent of the environmental catastrophe in the area, the Rural Municipalities of Westbourne and Lakeview readily agreed in 1937 and 1938 to lease some of their lands to Ducks Unlimited for twenty years.[100] After agreements were concluded with a number of private landowners in the vicinity, agents set to work restoring the marsh.[101] Their most important act was relatively simple: they dammed the drainage channels that had been dug with such a huge expenditure of time, money, and effort.[102] In some instances, local farmers had already tried to erect dams, and they were replaced by more permanent and effective structures.[103] Due to both the diligent work of Ducks Unlimited and increased rates of precipitation, the transformation was remarkable. Between 1940 and 1942, the agency claimed that the breeding population of ducks in Big Grass Marsh increased by 700 percent.[104] For a new factory, this was an astounding increase in productive capacity.

Ducks Unlimited offered a particular narrative of duck decline and recovery. The decline was portrayed as the result both of human actions such as overhunting and of natural processes such as drought-driven habitat reduction. This portrayal emphasized the agency's role in creating conditions conducive to a recovery in populations – and conducive to the loosened hunting regulations that only abundant waterfowl would justify. However, in a manner typical of inter-agency and inter-expert disagreements in the field of conservation, it was not a narrative with which everyone agreed.[105] The early 1940s saw matters come to a head as waterfowl experts disputed DU's optimistic assessment of waterfowl recovery. The American government's fish and wildlife officials were greatly concerned with DU's claims.[106] Alfred Hochbaum, a young waterfowl biologist and director of a research station situated southeast of Big Grass Marsh along the shore of Lake Manitoba, was similarly upset and became involved in personal confrontations with DU officials over the matter.[107] And Aldo Leopold, prominent conservationist and mentor to Hochbaum, went so far as to withdraw his support for the organization.[108]

For Manitoba farmers at least as concerned with the crops on their fields as the ducks on the wing, it was DU's wholly positive portrayal of increased duck numbers that generated friction. The Manitoba Department of Mines and Natural Resources received letters from individuals who thought, in the words of farmer Dane MacCarthy, they had been eaten "out of house and home" by waterfowl. MacCarthy went on to express frustration that Ducks Unlimited offered no compensation to farmers who suffered crop losses as a result of increased waterfowl populations.[109] The amount of grain lost could be quite significant. For instance, in 1946, farmer Fred Bintz reported a loss of eight hundred bushels of barley, valued at $560.[110] As Member of the Legislative Assembly Robert Hawkins explained to Minister of Mines and Natural Resources J.S. McDiarmid, farmers were convinced that, as a result of the work of Ducks Unlimited, "the ducks have increased to such extent as to cause increasing loss to farmers."[111]

Despite the doubts of other conservationists and the hostility of some Manitoba farmers, DU's publicity efforts contributed to the development of another way of thinking about wetlands. Manitoban E.S. Russenholt's description of his personal transformation from drainage advocate to wetland conservationist reflects this. Russenholt had a diverse working life. He worked for various agencies, including Manitoba Hydro as well as Ducks Unlimited (where his responsibilities included the drawing of cartoon publicity materials), achieved a measure of local renown as a television weather forecaster, authored numerous historical works as well as some children's fiction, and eventually retired to a farm just west of Winnipeg. Russenholt spent part of his boyhood working alongside his father and older brother to improve a homestead in Manitoba's Swan River region, west of Lake Manitoba and toward the northern edge of agriculture in the province. In a written remembrance, Russenholt recorded how in this area of the province "settlers regarded trees and water as the curse of the country." He recalled his boyish joy "when the government of the day arranged for big contracting outfits to come to the valley and dig a network of ditches to drain the water off the face of the land." But he also recalled his personal devastation a few years later when the drained land "shrivelled up under years of drought."[112] His perspective shifted further as he began to connect the changes wrought by farmers such as himself to negative outcomes such as diminished populations of wildlife and waterfowl. Russenholt was describing his personal experience of the intellectual shift that DU helped to promote among the population at large. Certainly, not everyone underwent such a dramatic about-face, nor was DU the only factor in changed ways of thinking. Nevertheless, even individuals more concerned with crops than ducks such as Dane MacCarthy and Fred Bintz (both of whom connected increased waterfowl populations to wetland restoration efforts) shared in the improved understanding of the functions of wetlands that Ducks Unlimited promoted.

Ducks Unlimited was created to ensure the availability of waterfowl for sportsmen, but its success depended on its attention to the broader environmental conditions (such as the existence of suitable wetland habitat) that favoured waterfowl as well as its ability to convince people that the concerns of its membership were in line with the broader public interest. Although the agency encountered resistance on various fronts, it transformed the physical landscape in Big Grass Marsh and worked toward a general shift in the perception of wet areas. As Manitobans became aware of the multiple functions of wetlands, they became less likely to dismiss them as nothing more than submerged farmland. And DU's influence was not limited to those individuals who would benefit personally from the increased waterfowl numbers the agency claimed would result from improved habitat. Ducks Unlimited contributed to a rapid renewal of Big Grass Marsh and to a gradual evolution in how all wetlands were perceived.

## INSTITUTIONAL INERTIA

In the mid-1930s, provincial government employee C.H. Atwood undertook a study of the water resources of Manitoba. Released as a publication of the Economic Survey Board of Manitoba, an agency created to help rationalize provincial administration amid the hardships of the Great Depression, the report addressed a wide range of issues related to water management, from urban water supply to hydroelectric development. It included an argument for a more holistic approach to the management of surface water, one that recognized the multiple purposes wetlands served and the significant challenges to managing such complex ecosystems. Atwood's *The Water Resources of Manitoba* was an early attempt to bridge the sort of new approaches evident in the work of Ducks Unlimited and before the International Joint Commission on the one hand and the standard public works approach to Manitoba drainage on the other.

But Atwood's report had little effect on government drainage administration. In the general preoccupation with drainage by district, with districts conceived in relation to drainage problems rather than watershed linkages, and in the persistent focus on drainage as a single problem rather than as one among a collection of environmental management issues to be addressed in relation to each other, drainage administration in Manitoba did not deviate substantially from a trajectory traceable in large measure to the 1895 Land Drainage Act. Perhaps the best illustration of this was M.A. Lyons's 1950 recommendation that watershed linkages be eliminated through the construction of double dyke drains to speed highland water through lowland areas. Even with respect to Drainage District No. 3, which was directly implicated in the Roseau River matter that helped to prompt progressive management of transboundary surface water, it was thought that double dyke drains were an appropriate solution to continued flooding. Such massive drains would trump natural flow patterns, splitting highland from lowland drainage and pre-empting the need for any sort of upstream/downstream negotiation or for further adaptation to the local environment. In District No. 3, as throughout the province, the sort of new perspectives evident in the work of Ducks Unlimited and before the International Joint Commission did not intersect with long-standing local debates over drainage.

How to explain this? Why did drainage, a key environmental management issue, remain more or less inured from changes in thinking that would seem to bear on it directly? There were three major reasons for this. The first factor was that in neither of the episodes involving the IJC and DU were the primary actors those most deeply involved in drainage. Federal rather than provincial officials negotiated a reasonable agreement to govern flood management along the Roseau River before the IJC. There seemed to be no need to reconcile long-standing provincial practices with the progressive federal approach. Furthermore, jurisdictional tensions between federal and provincial resource administrators persisted, and they had environmental as well as political consequences. Interestingly, C.H. Atwood was a federal government employee prior to his transfer to the province, and he had represented Canada's Department of the Interior at the Roseau River hearings.[113] Although the provincial government mandated him to undertake the study that resulted in his progressive report, it then failed to make any real policy changes on the basis of his recommendations. Not even the transfer of personnel could guarantee the transfer of progressive modes of thought.

Part of Ducks Unlimited's approach was to appeal directly to provincial officials who were, after 1930, responsible for administering the natural resources of the province. Building on the privileged social position possessed by many who were involved in DU at an early stage, waterfowl activists reached out to high-level administrators in the Manitoba government, with particular attention to the recently created Department of Mines and Natural Resources. Atwood was the department's first deputy minister, having taken up the post on his transfer from the federal government.[114] He was also one of DU's enthusiastic supporters and worked to promote conservation within his department. But drainage in this period remained a responsibility not of the Department of Mines and Natural Resources but of the Department of Public Works. Early on, the government decision to manage large-scale drainage under Public Works reflected a particular approach to the issue: surface water was an acute problem to be resolved permanently, not a chronic condition to be managed by farmers and government and certainly not an ecosystem to be nurtured. In the mid-twentieth century, as other ways of assessing surface water emerged, early administrative structures became barriers to progressive environmental management. Institutional mechanisms supporting drainage remained embedded in the practices of the Department of Public Works and were not reconciled with emerging conservationist approaches manifest particularly in the Department of Mines and Natural Resources. Even as one department came to appreciate the value of wetlands, another department continued its long-running efforts to eliminate any surface water that interfered with agriculture. Environmental administration in Manitoba operated on multiple tracks that failed to intersect.

Taken in sum, the various institutional barriers to the adoption of progressive surface water management, whether intergovernmental or interdepartmental divisions of responsibility, reflect how jurisdictional issues can compound the challenges of environmental administration. In one of the founding works of Canadian environmental history, historian Janet Foster argued that Canadian federal officials worked for wildlife preservation in part by exercising backroom influence on their associates, both those in government and those in prominent public positions.[115] With Manitoba surface water, the situation seems to have been quite different. Where federal officials found avenues through which to lobby for protective wildlife legislation, there were seemingly insurmountable institutional roadblocks to the adoption of progressive surface water management in Manitoba.

A second factor preventing the application of new environmental thinking to the long-standing problem of drainage was the fundamental connection between drainage and agriculture. Southern Manitoba, like the southern portions of Alberta and Saskatchewan, had been conceived by

Canadian administrators as an agricultural region. It was managed accordingly from an early date. Although the provincial economy had become more diversified over time, the rise of mining, hydroelectric development, and manufacturing had not displaced agriculture from its central role as the provincial economic motor.[116] As the province relied on agriculture, so did it rely on drainage. Manitoba's drained lands were both the most extensively cultivated and the most productive per acre.[117] Because of the connection between drainage and agriculture, the political and economic stakes were high for any administrative decision bearing on surface water – certainly far higher than they were for decisions bearing on wildlife management. This might have made it more difficult to enact policy changes, even once it was recognized that they were necessary.

The situation was further complicated by the array of challenges bearing on North American agriculture in the mid-twentieth century. Since the early 1920s, producers had been under pressure to expand their acreage and acquire expensive machinery if they were to achieve the economies of scale necessary to farm survival.[118] In Manitoba, the situation was conducive to continued efforts at drainage, as hard-pressed farmers sought to extract as much income as possible from their lands. Plowing around the wetlands that dotted some fields or tolerating intermittent flooding no longer seemed to be viable, particularly given the difficulties of performing such manoeuvres with large, modern agricultural equipment. Increased pressures on farmers translated into increased pressures on the land. The drive to achieve farm efficiency amounted to new fuel for the drainage enterprise.

Ducks Unlimited was made up largely of individuals who had found professional success, often in the business world. Agency officials could identify with farmers' drive for efficiency. Indeed, Ducks Unlimited promoted wetlands conservation through a logic that was consistent with farmers' economic approach. Beyond the focus on ducks instead of crops, the difference was that DU officials applied the idea of efficiency to the regional landscape rather than to the individual farm. In this view, farmers could maximize production on their acreages, through drainage if necessary, even while supporting the preservation or restoration of wetlands located elsewhere. And given that DU advocates were interested primarily in ducks, not wetlands, the ready food source provided to waterfowl by farm fields represented a real benefit. Ultimately, there was little acknowledgment that the wet prairie that farmers drained away and the wetlands that DU sought to create were in fact ecologically similar landscapes, at least insofar as both could support migratory waterfowl. Ducks Unlimited promoted an alternative view of prairie surface water in which wet areas played a valuable role in waterfowl life cycles. But this view was not in conflict with the agricultural vision for the region. It was certainly not a coincidence that Duck Factory No. 1 was located in an area that, even when drained, was not productive farmland.[119] DU worked within the human as well as the environmental landscape of the province, promoting a vision of the wet prairie that could support more waterfowl without entailing a sacrifice of farm acreage. Ultimately, it is not that Ducks Un-limited failed to change the view of wetlands held by Manitoba farmers or government administrations; rather, the agency tailored its message so that there seemed to be no contradiction between agricultural production and wetland preservation.

There was at least one other major reason for institutional inertia in Manitoba's drained landscape. By the mid-twentieth century, the drainage system in Manitoba represented a substantial investment of time, money, and effort. This was an infrastructure that focused on lands in need of drainage, not on the larger watersheds in which they were embedded. The spatial units created by the drainage infrastructure became the lens through which the provincial environment was perceived, at least for many govern-ment administrators. The physical infrastructure of drainage, though not always successful at keeping the water flowing, had tenure in part for the simple reason that it was already in place. Ultimately, it became the physical anchor for an outdated perception of the provincial environment.

The concept of technological momentum helps to make sense of the situation. Through examination of the relationship between technology and society, historian of science Thomas Hughes has argued that newer technological systems are more open to social influences, whereas older systems tend to be less vulnerable to forces of change.[120] Manitoba engineer F.E. Umphrey explained the local situation in typically awkward prose: "Watersheds are important only from an Engineering point of view for the purpose of the comprehensive design of an efficient and sufficient system." He went on to explain how, at this late date, they would not necessarily play a helpful role in resolving the province's drainage prob-lems.[121] From this view, the problem was less with the watershed idea itself than with the fact that it came so late to Manitoba. The province missed its chance to introduce watershed management in 1895 with the passage of drainage legislation that did not incorporate the watershed idea. Through the construction of substantial physical infrastructure, developments since that time had set the province on a course that led away from watershed management. Once oriented to the administration of the drainage system and occupied with the various tasks associated with it, the Department of Public Works did not seriously entertain other ideas about how the surface water situation could be managed. The drainage infrastructure pointed the province in a certain direction, even when the drains themselves failed to adequately transport surface water.

## CONCLUSION

The dominion government became involved in the Roseau River situation because it was an international concern. Since 1870, Manitoba and Ottawa had often pulled against each other on matters related to land drainage. Even arrangements made cooperatively sometimes degenerated into antagonism through problems of implementation. With regard to the Roseau River, federal and provincial officials came together to present as convincing a case as possible in the hope of countering American claims. Despite the antagonism suggested by Canadian concern to present a united front, the international context fostered a more constructive discussion of drainage that moved beyond upstream/downstream conflict. Both Canada and the United States appreciated the need to negotiate a fair agreement, as advantages won along one river would amount to losses along another. The result was a more sophisticated approach to the challenges inherent in water management than had emerged from the Manitoba conflict between highlanders and lowlanders.

In Manitoba, wetland drainage was a provincial story: early drainage was a provincial initiative; municipalities took over with provincial encouragement and support; and Manitoba's 1895 Land Drainage Act established a basic legislative structure for large-scale drainage works. The roots of wetland restoration, however, require explanations that address international factors: conservation as a mode of thought; the drought of the 1930s as an experience shared by many; and the transnational efforts of Ducks Unlimited. Clearly, locals always had big dreams for Big Grass Marsh. As developing concerns with conservation intersected with the drought of the 1930s, the marsh became significant to people who lived far from it. Ducks Unlimited helped to change the perception of wetlands through its publicity efforts by making apparent the role of marshes in the international waterfowl life cycle. Manitobans and others came to appreciate the advantages of the conservation and restoration of certain wet areas.

Along the Roseau River and in Big Grass Marsh, thinking about Manitoba's water became more sophisticated due to the involvement of wider contexts and the introduction of new modes of thought. Both international influences and conservation ideas came into play due to bioregional linkages (floodwater flow patterns on the one hand and waterfowl migration routes on the other) that spanned geopolitical borders. But discussions of transboundary water management and marsh restoration did not directly intersect with local debates over drainage. With watershed management, as with prairie settlement, the federal government had the advantage of working at a higher level of abstraction, one that made it easier to envision a diverse region divided into a settlement grid or to embrace the ideal of watershed management despite decades of individualistic land policy. The provincial government lacked the advantage of distance and thus remained mired in the practicalities of, first, making individual farms viable in an ecological commons and, later, reconciling the watershed ideal to the private property landscape it had worked so hard to establish. Ducks Unlimited was in no way constrained by the jurisdictional arrangements that bound both the province and the dominion; indeed, the organization demonstrated a remarkable capacity for uniting big dreams and practical change. But while recognizing this it is also important to consider the limitations of what the agency achieved. Even when Ducks Unlimited succeeded in reaching the rural farmer who was most interested in drainage, there was little attempt to reconcile the select wetlands identified as valuable waterfowl habitat with the wet patches that continued to be perceived as barriers to agriculture, though the hydrological differences among them were largely matters of scale. In the 1930s, even as provincial bureaucrats and politicians came out in favour of habitat preservation, the institutional mechanisms that supported drainage remained embedded in the structure of the provincial state. And few mechanisms to protect wetlands had yet been introduced. In the larger context of the history of drainage in Manitoba, the events in Big Grass Marsh and along the Roseau River are most remarkable for what they failed to do: substantially alter the thinking of those most concerned with the question of drainage.

# 5

# Permanence, Maintenance, and Change: Watershed Management in Manitoba

In October 1975, Manitoba government employee William R. Newton appeared before the eighty-ninth annual congress of the Engineering Institute of Canada to explain changes to surface water management in the province. Despite his focus on recent developments, Newton prefaced his presentation with a discussion of the agricultural system that prevailed across the Prairies, linking many enduring problems to how "in our planning we have been more concerned with geometry than geography."[1] This was a reference, and certainly not an oblique one for the expert audience he was addressing, to the grid-based township survey that had guided early prairie settlement. In effect, he was telling a roomful of engineers, masters of human-engineered efficiency, that prairie agriculture might have been more successful under a less engineered, more environmentally attuned approach to settlement.

This chapter explores how governments and people changed their minds and their practices in relation to the wet prairie. Early on, Manitobans conceived drainage projects as permanent improvements to the agricultural landscape. As inadequately maintained ditches slumped in on themselves, Manitobans began to question their assumptions about the permanence of the drainage infrastructure. But even once the need for drain maintenance was understood, confusion between provincial and municipal governments over responsibility for upkeep meant that necessary work often went undone. The situation was exacerbated by snowballing problems with drainage funding, at the core of which were awkward arrangements established by the 1895 Land Drainage Act. The environmental and financial problems that bedevilled drainage in Manitoba were compounded by the widespread agricultural and economic crises of the 1930s. The lowlands were in a desperate state. However, given enduring antagonism between highlanders and lowlanders, a lowland predicament was simply not suffi-cient to catalyze a broad change in provincial surface water management.

Although lowlanders had long argued for highland involvement in drainage funding, highlanders had much preferred to remain aloof from a set of problems they did not recognize as their own. But by the middle of the twentieth century, highlanders could not deny the land erosion under way in highland areas. As residents grasped the severity of the situation and cast about for possible remedies, they were prompted to rethink not only how they managed their own lands but also how they related to their neighbours and the provincial government. Some highlanders began work-ing with some lowlanders in a bid to establish environmental management arrangements that would address in tandem the challenges of erosion up above and flooding down below. In this way, the highlander/lowlander impasse that had long forestalled innovative surface water management was finally circumvented.

But this new perspective did not eliminate all barriers to the develop-ment of new surface water strategies. In the late 1950s, reflecting the dif-ficulties of modifying established practices, a period of administrative confusion ensued within the provincial government itself. Was drainage more appropriately the responsibility of the Department of Public Works or of the Department of Agriculture? Or did it require the development of a new entity with a new set of responsibilities, such as a Department of Conservation or a Manitoba Water Commission? Answering these ques-tions was rendered more difficult by rural economic and demographic changes. New opportunities to secure funding from the federal government were not sufficient to compensate for all of this. And there were lingering difficulties stemming from the historical relationship between the prov-incial government and Manitoba residents. After decades of protracted disputes over how flooding should be abated and who should pay for the necessary measures, surface water administration was as much dispute management as water management. The frustration and distrust built up through decades of failed government efforts served as barriers to the co-operation essential for any new initiative. Ultimately, contention over drainage had become an environmental problem in its own right.

PERMANENCE, MAINTENANCE, AND MUNICIPALITIES

The history of catastrophic flooding along the Red River surely must be one of Canada's most compelling illustrations of the hubris of European settlement. Early settlers in the region were devastated by floods on numerous occasions and became accustomed to coping with high levels of water. But as formal governments were established and the population increased (and despite warnings from some who saw ominous signs in the region's history), settlement patterns were not modified to accommodate the risk of flooding. The transcontinental railway provided perhaps the best illustration of government and settler unwillingness to adapt ambitious plans to local environmental conditions. Shortly after Confederation, Chief Engineer Sandford Fleming was assigned the task of determining the best route across the prairie landscape. After extensive study of the local geography, he recommended that the rail line be run not through the fledgling metropolis of Winnipeg but through the more northerly, less flood-prone (though certainly not flood-immune) town of Selkirk.[2] "If, without due consideration, or regardless of the local experience which has been gained by many now living, we were to carry the Railway across Red River anywhere in the district subject to inundation," Fleming warned, "we might any year find a dozen miles of the line for a month or more submerged, the bridges and approaches swept away, and traffic stopped until the whole be restored."[3] Given the importance of the transcontinental railway to the development of the northwest, such an outcome could have dire consequences.

But Fleming's warning did not suit the ambitions of the Winnipeg business community. Powerful local capitalists thought their fortunes depended on securing the railway, and they launched a substantial campaign in favour of the Winnipeg route.[4] This class-based lobby effort coincided with what historian J.M. Bumsted has called the province's "folk wisdom," a general belief that "somehow the extent of human development along the river banks had reduced the flood danger."[5] Some believed that widening the river channel had lessened the risk.[6] Although much uncertainty surrounded the relationship between drainage and flooding, others maintained that agricultural drainage would protect against catastrophic flooding.[7] Governments were amenable to efforts to downplay the risk, largely because they did not want to deter potential immigrants. In light of a rough consensus that flooding would not be a major problem, the railway was built through Winnipeg. Riverbank property that was most vulnerable to flooding became the city's most desirable real estate.[8] The province's high ridges, which had once provided refuge in times of flooding, were used for other purposes, including the construction of a prison at Stony Mountain.[9] Flooding had afflicted early settlement in the area, but Manitobans and their governments did not modify their land use patterns to take account of this history.

In 1950, a large-scale flood along the Red River ravaged the provincial south and Winnipeg.[10] After seventy years without an inundation of comparable magnitude, and in light of a dogged unwillingness to modify land use practices to accommodate flooding, Manitoba was "singularly unprepared" to cope with this flood. The province was "almost totally lacking the most elementary structural protection – such as dykes – and emergency planning measures."[11] In all, 1,664 square kilometres (642 square miles) of land north of the international border were flooded.[12] Only extreme effort sustained over weeks along the hastily constructed sandbag dykes that snaked through the city saved parts of Winnipeg from inundation. In the wake of the flood, the provincial and federal governments cooperated on large-scale projects intended to guard against future floods. The Red River Floodway, a flood protection channel designed to divert Red River floodwaters around Winnipeg, was one of these projects. This was among the largest earth-moving projects in human history to that time, and the size of the project was an indication of the extent of environmental maladjustment in the region.

How did these developments relate to the experiences of Manitobans who relied on agricultural drainage? From a settler perspective, the problems of inadequate drainage and catastrophic flooding looked quite different from each other. In the one case, water took too long to make its way to the rivers; in the other, water rushed out over the riverbanks. But from another angle, surface water flooding is simply catastrophic flooding writ small, with a similar sort of environmental hubris at its root. Any apparent improvement in environmental conditions was regarded as a lasting change. Environmental problems were thought to be relatively easily resolved. Both government officials and local residents were only too eager to believe that the landscape was entirely suited to the ways they wanted to use it. The same overreaching optimism that helped to drive the rail line through Winnipeg defined how settlers and their governments perceived the challenges they confronted in what they envisioned as an agricultural landscape. Scholars such as Hugh Prince, John Thompson, and Ann Vileisis have made it clear that settlers and governments across

North America saw drainage as a means of ensuring long-term agricultural prosperity.[13] The situation was similar in Manitoba. Surface water flooding might not take care of itself, but optimistic residents of early Manitoba assumed that solving the problem would be a relatively simple and wholly successful endeavour.

There was also a more specific reason for the belief, widespread in Manitoba, that drainage would permanently resolve any problem with surface water flooding. Under the terms of Confederation in 1867, the dominion government assumed some of the debt accumulated by Canada, Nova Scotia, and New Brunswick in the construction of essential public works. When Manitoba entered the union a few years later, some means of extending equally generous terms to the new province had to be found. As no substantial infrastructure had been built in the Red River settlement, there was no debt for the dominion to assume. Consequently, the dominion government promised funds for capital improvements in Manitoba, provided that these projects would constitute an enduring asset to the province.

Confronted with a barren provincial treasury on the one hand and frequently inundated agricultural lands on the other, Manitoba politicians emphasized the permanent character of drainage works in an effort to leverage these funds. In a letter to the receiver general of Canada, the Manitoba premier and treasurer, John Norquay, sought the substantial sum of $100,000[14] for drainage works that had become "a matter of necessity for the welfare of the settlement of the Province." As they were of "a permanent character," he explained, they were rightly a charge against the capital account rather than the current expenditures of the province.[15] In this view, drainage was unique, not only in its significance to the province, but also in its nature as a public investment: "While roads and railroads will wear out, bridges will be swept away, and buildings wear out and deteriorate, the ditches if properly located and made in soil such as that of which Manitoba is chiefly composed, will continue to enlarge from year to year."[16] Bridges and public buildings along with drains were necessary to provincial progress, but only drains were resistant to deterioration, even likely to improve over time. The widespread North American understanding of drainage works as permanent improvements was re-enforced in Manitoba as part of an effort to secure funding from the dominion government.

The reality of open drains in a dynamic region was very different. Although the nearly exclusive use of surface ditches for drainage in Manitoba substantially reduced capital costs, it also created a greater need for ongoing maintenance. Surface drains were vulnerable to processes of erosion and sedimentation that could significantly impede their capacity to transport water quickly and effectively.[17] By the mid-1910s, many low-landers were convinced that land use changes in the highlands increased surface water flows into the lowlands, with the result that the existing drains were overwhelmed. But they also realized that, in the absence of regular maintenance, the drains themselves deteriorated rapidly. Norquay was simply wrong in his claim that drains would improve over time. Indeed, his politically motivated assertion was as unreasonable as the idea that settlement had somehow diminished the risk of catastrophic flooding along the Red River. Drainage changed the landscape but failed to resolve the basic fact of environmental variability, which was ultimately to blame for much continued flooding.[18]

Open ditches proved to be particularly problematic in light of how agricultural intensification affected the regional environment. Although the 1930s brought difficulties with regional agriculture to the fore, the problem of topsoil loss was evident across the Prairies from relatively early on. In 1901, the annual report of Manitoba's Department of Agriculture and Immigration explained that "the prairies, which had for years been storing up humus, and whose particles of soil were knit together by the fibrous roots of grasses, have in some districts become moving drifts of soil.[19] Topsoil blown by the wind collected in low areas such as drains. In 1922, engineers estimated that the capacity of some of Manitoba's early drains was already diminished by as much as 40 percent to 50 percent.[20] Matters worsened over subsequent years, with the winds of 1925-26 so "disastrous" as to result in drains "being blocked for half a mile at a stretch."[21] A problem that was building at the turn of the century was a crisis by the 1930s, particularly as dry periods exacerbated drain degradation.[22] Farmers in Manitoba's highlands watched as dry winds stripped fertile topsoil from their fields, while farmers in the lowlands observed with equal dismay as the material lodged itself in their ditches.

The topsoil that was so valuable on the fields was nothing but a problem in the ditches. Clogged drains caused water to back up over the fields, but "the cost of cleaning out drains after every wind storm [was] heavy."[23] The material that ended up in the ditches had lost none of its fertility, though it was no longer available to farmers. Weeds were quick to colonize the silted ditches that, in Manitoba's variable climate, were often dry for part of the growing season. This weed growth further anchored the blockages and complicated drain maintenance. By 1930, the willows in the drains of Drainage District No. 2 were nothing less than "a menace that must be dealt with."[24] As farmers in the early twentieth century became increasingly concerned with weed growth, and as governments established policies designed to encourage eradication, unmaintained drains became notorious as vectors of weed dispersal.[25] In 1947, a machine became available to apply the newly developed herbicide 2,4-D to weed-clogged ditches.[26] Although neither the extreme toxicity of this chemical nor the particular dangers of applying it directly to a waterway were widely recognized at the time, the action still reflects the desperation of those relying on clear waterways to drain their fields. Drains that failed to transport water adequately because of siltation proved to be ideally suited to the diffusion of unwanted plants.[27]

Drain degradation was also worsened by farmers' efforts to adapt an infrastructure tailored to wet years to circumstances in dry periods. Some farmers deliberately obstructed drains. Erecting a small dam was an easy way to create a convenient source of water for specialty crops or livestock. This practice became especially prevalent in areas where groundwater resources were inadequate or unsuitable, such as throughout much of the clay-soiled Red River valley.[28] Even if farmers removed the dams during periods of high water, pooling behind obstructions would induce deposition of sediment that would remain suspended in flowing water. Dry years contributed to infrastructure degradation that would have catastrophic potential when wet years returned.

Obstructed ditches disrupted water flows, causing problems for those who depended on the drains. But flows of cash between the province, the municipality, and the drainage district were every bit as problematic and came into sharp focus over the question of maintenance. Drainage figured in the development of the municipal system in Manitoba as the need to address surface water flooding helped to drive the extension of government authority. Over the years, municipal governments acquired significance as they played a role in local affairs and cooperated with the Manitoba government on larger projects of local consequence. In the years before The Land Drainage Act, drains were funded as entirely municipal, entirely provincial, or joint undertakings, depending largely on size and location. Under The Municipal Act, maintenance of drains, as well as of other local infrastructure such as roads and bridges, was the responsibility of the municipality in which they were located.[29] All of this was funded out of revenue generated through municipal taxes, with occasional assistance from the provincial government. After the 1895 passage of The Land Drainage Act, looking after drains became a much more onerous responsibility. Indeed, maintaining the infrastructure of a drainage district was an undertaking on an entirely different scale from the care of the small drains that had existed prior to 1895. Some municipal officials were not aware of their responsibility for drainage district drains, assuming that, as the province was responsible for construction, so was it responsible for upkeep. Others objected to the arrangement and, in protest, resisted spending their limited municipal resources on maintenance. Neither case was conducive to clearing out clogged drains.

The 1895 Land Drainage Act was designed to facilitate large-scale drain construction, but municipal governments remained fundamental to project financing. Drainage districts lacked independent administrative infra-structures and thus were unable to collect their own levies. The munici-pality became responsible for collecting the assessment from the residents of all drainage districts within its borders and for forwarding the money to the province. Drainage districts typically encompassed a number of municipalities. For instance, Drainage District No. 2 included portions of the Rural Municipalities of Morris, Roland, Macdonald, Dufferin, Grey, Portage la Prairie, and Cartier. Each municipality was responsible for col-lecting the drainage levy from the landowners in its portion of Drainage District No. 2 as well as for any other drainage district within its jurisdic-tion. Collecting the drainage taxes was a significant imposition, especially when a single municipality overlapped several drainage districts.

The provincial government also used municipalities to prop up drainage districts in financial ways, and this provoked dissatisfaction among mu-nicipal residents as well as municipal officials. Municipalities levied taxes on all residents to cover the costs of local services and infrastructure. The rate of taxation for which landowners were liable was determined partly in relation to the value of their property. The Land Drainage Act speci-fied that, for the length of time it took a drainage district to pay down the drainage debt, no adjustment could be made to the assessed value of the drained land.[30] As a result, after a few years, landowners benefiting from drainage were still paying municipal taxes as if their lands were waterlogged. Drained land worth $60 to $100 per acre was assessed as if it were still wet land worth $2 or $2.50.[31] Since drainage levies were to be paid over periods of twenty-five or thirty years, those in drainage districts were given what amounted to a long-term break on their municipal taxes.

The frozen municipal assessment was used as an incentive by govern-ment officials and drainage advocates as they sought to sell the idea of drainage district formation to skeptical landowners. But this incentive distorted the power of municipal officials to increase taxes. The Land Drainage Act imposed no restriction on rates levied on municipal residents outside drainage districts, but rates within drainage districts were restricted.

Municipal authorities had to decide whether to increase levies dramatically on those outside drainage districts or to curtail municipal infrastructure projects. The larger the portion of a municipality included in drainage districts, the more difficult the predicament of local officials. As drain maintenance had to come out of municipal funds, those living outside drainage districts ended up contributing the largest proportion of the money to fund drain maintenance and other municipal undertakings. To municipal residents beyond drainage district boundaries, it seemed that those in drainage districts paid less but got more. That this situation "did not lend itself to the raising of money for the maintenance of ditches" was readily apparent to members of the Sullivan Commission by 1921.[32]

Even those who lived in drainage districts sometimes failed to appreciate or acknowledge that they were getting a break on their municipal taxes because even a low municipal tax could add up to a substantial sum when inflated by the drainage levy. Also, since drains were built relatively quickly but paid for over time, their effectiveness could be assessed long before the debt was retired. In many places, it soon became apparent that drains did not live up to local hopes. Unreasonable expectations were a problem, for surface water continued to muddy the fields of many who had antici-pated more profound environmental change. Even when drains appeared to be relatively successful, disappointment could ensue if the drained land turned out to be less productive than had been hoped. Inadequate main-tenance also left many lowlanders dissatisfied. Some drainage district residents withheld payment of the drainage tax out of anger and frustra-tion; others, especially in difficult economic times, simply could not pay their bills.[33] Since the municipality collected the drainage levy for the province, landowners who did not pay became a municipal problem.

These awkward administrative and financial relations among the prov-ince, the municipalities, and the drainage districts laid the foundation for a crisis in drainage financing. The catalyst was the economic catastrophe of the Great Depression. When prices for agricultural products plummeted and farmers defaulted on their municipal taxes, many municipalities were unable to meet their financial obligations to the province.[34] In some areas, the burden of drainage costs contributed to farm abandonment, which only served to worsen the municipal position as the number of residents eligible for taxation dropped.[35] Between 1928 and 1935, the government was obliged to allow sixteen municipalities to postpone drainage payments. On 30 April 1935, municipalities owed the province almost $4 million in debt related to drainage projects – a substantial sum on a drainage infra-structure valued at approximately $7 million.[36] Adjustments were necessary if municipalities were to remain solvent. This was a challenge to the integrity of the drainage district system every bit as significant as the soil that clogged the ditches.

In 1935, the Finlayson Royal Commission was tasked with the investigation of flow interruptions, both of cash and of water. It recommended a sweeping reorganization of the administrative infrastructure for drainage. Drainage districts were to be renamed drainage maintenance districts, which reflected a fundamental reorientation from system expansion to infrastructure maintenance. A board of maintenance trustees was to be established for each drainage maintenance district, consisting of one representative from each municipality within the district. The board was to be granted full control over the maintenance of ditches and equipped with a staff to manage the work. It would have at its disposal funds supplied by the province in addition to money raised through an annual levy on residents of drainage maintenance districts. All boards were to be chaired by an individual appointed by the lieutenant governor in council.

The recommendation for enhanced provincial funding depended on one of the report's major assumptions: that the period of intensive drain construction in Manitoba was at an end. In the commissioners' assessment, the existing drainage system was more or less sufficient – or at least it would have been if properly maintained. Under the expectation that ongoing maintenance would be far more economical than new construction, it was anticipated that drainage costs were about to drop substantially. Future expenditures would thus be more in line with the available resources. This assumption served to moderate the financial burden to be assumed by the provincial government. None of this seemed to be unreasonable in the dry years of the 1930s. With less water in the ecosystem, there was little risk of flooding, and there were few tangible reminders of the inadequacies of the drainage infrastructure. And the provincial government was willing to go some distance toward making the new arrangements work. The newly appointed chair of the drainage maintenance boards, engineer F.E. Umphrey, set about establishing a new administrative infrastructure and cultivating goodwill among those who had so long been dissatisfied with drainage.

But the new arrangements turned on an unhelpful, even misleading, distinction between construction and maintenance. The Finlayson Commission assumed that, in contrast to the expense and effort involved in construction, maintenance implied moderate projects and modest expenditures. This was not always the case. Ongoing upkeep could be as costly as initial construction. Although by this time few believed (as

Norquay had argued) that drains were a permanent improvement to the provincial landscape, many still failed to recognize that drainage was an ongoing, adaptive endeavour driven by both anthropogenic and non-anthropogenic environmental changes. Expectations had changed, but they remained unreasonable.

The situation was exacerbated as, over the years, the Manitoba government interpreted the commission's report in ways that minimized the province's obligations.[37] The province capped its contributions to all drainage maintenance districts at $30,000. Although this amount was later raised to $40,000, any cap left the districts liable for heavy costs in particularly wet seasons.[38] During the late 1930s, a period that remained relatively dry, drainage expenditures fell well within the $30,000 allowed by the province. Indeed, a small surplus was accumulated and made available to any district in need of additional funds.[39] Maintenance boards chair Umphrey did what he could: assisting with the establishment of the drainage maintenance boards, managing the equipment to be used in maintenance, and establishing strong working relationships with municipal and provincial officials and the trustees of the drainage maintenance boards. Yet he knew full well that he was simply making good use of the calm before the inevitable storm. With inadequate government support, wetter years would surely bring trouble.[40]

In 1943, it started to rain. Higher water quickly exposed the "many instances of errors in judgment and neglect in maintenance" over the years since the Finlayson Commission.[41] More significant still were the financial difficulties. As the province remained unwilling to tie the amount of its contribution to annual expenditures, drainage maintenance boards were stretched between an inflexible provincial government and a dynamic wet prairie landscape. Some boards began asking municipalities for financial assistance with necessary work beyond what was covered by the provincial grant. Still, it was impossible to fund sufficient work to offset entirely the higher water levels. Once again, lands were flooded, and Manitobans were unhappy.

On 27 June 1944, concerned municipalities formed the Union of Municipal Drainage Maintenance Districts (UMDMD). Twenty-four of the twenty-seven municipalities in drainage maintenance districts were members by 1947.[42] The UMDMD demanded a government investigation into the drainage situation. Realizing that the problem was not likely to solve itself and faced with new public pressure, the government increased funding to the drainage maintenance districts and began to consider ways in which additional resources would be made available in times of crisis.[43]

Although the province was unwilling to establish another full-scale royal commission, it acknowledged that action was necessary both to address the current situation and to prevent similar crises in the future.

By order in council of 14 January 1947, civil engineer M.A. Lyons was appointed to investigate the drainage situation.[44] Lyons's review was far more comprehensive than the government had envisaged. It included interviews with municipal officials and farmers and inspection trips to each drainage maintenance district.[45] Lyons operated in a context very different from that of the Finlayson Commission. The Finlayson report addressed a concrete financial problem in a specific environmental context: a municipal debt crisis in a dry period. By 1947, more rain had exposed the government's failure to implement some of Finlayson's major recommendations and highlighted some of the report's limitations. This relatively recent increase in rates of precipitation provided an especially vivid illustration of the dynamic nature of the Manitoba environment. Lyons determined that it was necessary to develop drainage arrangements predicated on environmental variability.

Accommodating a dynamic environment meant tailoring funding to environmental conditions. Environmental changes both anthropogenic (continuing road building and land clearing) and non-anthropogenic (ongoing swings between wet and dry periods) could require substantial modifications to existing infrastructure to cope with altered drainage patterns.[46] Effectively, there was no useful distinction to be made between new construction and system maintenance. Furthermore, Lyons argued that there should be no fixed cap on government contributions. While $90,000 or $100,000 was estimated as a reasonable average annual government contribution, a key aspect of Lyons's recommendation was that the final figure would rise and fall entirely in relation to actual costs.[47] Drainage funding would not be permanently fixed. It would be adjusted on an ongoing basis, in relation to prevailing conditions. Such environmental realism was in stark contrast to the unrealistic optimism that animated early drainage efforts.

Lyons's recommendations, based on his expert study, were in tune with what the chair of the drainage maintenance boards advocated on the basis of his practical experience with drainage administration. In a November 1945 presentation at the annual meeting of the Union of Municipal Drainage Districts, Umphrey noted how wet periods inevitably resulted in a mismatch between available funds and necessary work. Flood-vulnerable Manitobans wanted work performed when environmentally necessary, not as funds became available. Recognizing this, Umphrey argued that the most effective means of ensuring that financing did not delay construction was through the provision of what he termed a "line of credit." This meant a standing agreement by which funds would be made available, presumably by the government, for pressing work. Umphrey explained his proposed arrangement by comparing it to assistance with seed grain or fodder for the needy farmer, explaining that, in drainage as in any business, "we MUST have the funds to do work as and when required and NOT as and when funds are available." A line of credit would provide an alternative to what Umphrey described as the current "cash and carry system," under which necessary maintenance would not be performed if sufficient funds were not already at hand in the district in question. Ultimately, it was the environment that would define funding rates, with the amount of money devoted annually to drainage contingent on actual conditions. Taken in sum over a period of years, government funding would amount to "the average of the [drainage] expenditure over a seasonal cycle including so many dry years plus so many wet years."[48]

Both Umphrey and Lyons were arguing that public funds should be used to moderate the consequences of environmental variability. Umphrey's description makes it especially clear that the need was not only for more government money but also for a new system through which funds would be made available. More intensive state involvement was necessary since funding levels should be adjusted on an ongoing basis to reflect environmental conditions. This expanded role for the Manitoba government in drainage financing was in keeping with contemporary thinking in the post–World War II era as politicians and administrators grappled with Keynesian economic principles. Under Keynesianism, public funds would be used to moderate the consequences of economic downturns. Although subject to change over time and from administration to administration, economic policy derived in some measure from Keynesianism was influential in Canada from the mid-twentieth century onward.[49] There is a rough parallel between government spending under Keynesianism (public funds employed to protect the well-being of Canadians in times of economic downturn) and what both Lyons and Umphrey proposed for drainage funding (public funds employed to shore up the drainage infrastructure in times of increased precipitation). In both cases, state action was to safeguard citizens from the worst effects of inevitable swings, whether economic or environmental.

Over time, it became apparent that, despite the optimism of early settlers and governments alike, coping with surface water would be a long-term challenge, one that involved the management of human expectations as well as environmental conditions. As the construction of the Red River Floodway signalled a greater appreciation of the conditions of settlement at the forks of the Red and Assiniboine Rivers (even as it amounted to an effort at environmental alteration rather than adaptation), so Lyons's recommendations reflected a more realistic assessment of environmental variability, one of the basic challenges to agriculture in Manitoba. Distilled to its essence, his key insight was that surface water problems inevitably would persist, even in the drained landscape.

## SOIL EROSION AND WATERSHED MANAGEMENT

There had been early hints that Lyons's report would be innovative not just in relation to environmental variability but also with regard to management by watershed. A brief from the Union of Manitoba Drainage Maintenance Districts that had prompted the government to commission Lyons's inquiry specifically referred to the importance of placing the foreign water problem in its watershed context.[50] At his own request, Lyons was charged with exploring how surface water problems related to environmental management broadly conceived.[51] His appointment, which referred to an investigation of "drainage, water conservation, and possible irrigation projects in Manitoba," gave him a mandate to think about watersheds.[52] When juxtaposed with the cooperative approach exhibited by Canadian and American officials before the International Joint Commission hearings in the 1920s and the conservation activities of Ducks Unlimited in the 1930s, it is clear that Manitoba drainage administrators were not particularly innovative in these decades. However, by the time Lyons began his investigation in the mid-1940s, and likely due in part to the continued dissemination of ideas that had early found purchase in Manitoba along the Roseau River and in Big Grass Marsh, some government officials and Manitoba settlers favoured rethinking government policy in a more holistic way rather than dealing with drainage in isolation.[53] There was greater appreciation for the wet prairie as an ecosystem of value in itself rather than an agricultural problem. Ideas about wetlands and drainage were certainly not as changeable as environmental conditions in the wet prairie, but neither were they by any means permanently fixed.

But despite both his initial orientation and the changed public context, Lyons's report argued that the watershed concept would be of little practical use in contemporary adjustments to the drainage infrastructure. Although he continued to see the appeal of apportioning drainage costs on a watershed basis and noted that the first drainage investigation under J.G. Sullivan proposed something along these lines, Lyons recognized the contentious conceptual realm in which he operated, and he concluded that a broad consensus on this approach was simply out of reach.[54] The deeply entrenched and fundamentally opposed ideas of highlanders and lowlanders over the idea of watershed management, not any weakness in the concept itself, was the key factor in his decision not to recommend that the provincial government adopt watershed-based drainage financing. While Sullivan's 1921 report was inspired by idealism, Lyon's report was defined by realism, and management by watershed did not seem to be a realistic possibility at the time.

Ultimately, it was not water but soil that would eventually sweep away highlander resistance to watershed management. Within the province as around the world, agricultural cultivation had altered soil quality, reducing the organic component. Degraded soils were finer in grain and thus more likely to crust or drift. In combination with dry, windy weather, it was the breakdown of soil structures that resulted in the dust clouds noted across the North American Great Plains in the 1930s.[55] Although governments and farmers had long been aware of the risk to the soil, this difficult decade prompted new efforts to find solutions to what had become a chronic problem.

Like Alberta and Saskatchewan, Manitoba suffered from soil erosion by wind.[56] But the province's soup bowl topography and higher average precipitation also contributed to a significant problem with soil erosion by surface water. Runoff swept degraded soils from upland areas, with consequences for field fertility and, in steeper areas along the Manitoba Escarpment, slope stability.[57] The transport of material from upper to lower watersheds was not a new problem. Indeed, it contributed to the infilling that had long impaired the drainage infrastructure in the lowlands. Both the greater attention to soil conditions that grew out of the crisis in the 1930s and the increased runoff that followed the end of the drought contributed to a change in perception. What had once been conceptualized primarily as the problem of sedimentation in drains was increasingly viewed as a problem of erosion. By 1950, it was estimated that almost 2 million acres (809,000 hectares) of Manitoba's agricultural land was subject to considerable erosion.[58] Over the following years, the matter was of concern not only to lowlanders vulnerable to inundation but also to highlanders increasingly worried about soil loss.

For municipalities southwest of Lake Manitoba, 1956 was a particularly difficult year. There was significant erosion in the highland Rural Municipalities of Lansdowne and Rosedale and severe flooding in the neighbouring Rural Municipality of Westbourne. In light of this shared experience of hardship, representatives from these and other afflicted areas began to consider whether a watershed-based approach might lead to more satisfactory environmental management. Through meetings over the summer and fall, the Riding Mountain-Whitemud River Watershed Committee was developed to promote the watershed idea among local residents and to pressure the provincial government to develop appropriate institutional supports.[59] The activities of the committee were consistent with a tradition of locally based lobby groups (such as the Red River Valley Drainage and Improvement Association and the Union of Municipal Drainage Maintenance Districts) that in the past had succeeded in influencing the Manitoba government. However, this committee was different because it united highland areas suffering from erosion and lowland areas suffering from flooding.

The provincial government was receptive to the committee's concerns, and a series of consultations culminated in the 1958 passage of The Watershed and Soil Conservation Authorities Act. The Manitoba statute provided for the creation of a local infrastructure to "promote the conservation and control of the water resources and other related resources" and recognized the value of defining districts on a watershed basis.[60] As explained by W.R. Newton, director of operations for Manitoba Water Resources, the major impetus behind the 1958 legislation was the recognition that Manitoba's watersheds were "sick and getting sicker," with erosion and flooding as related afflictions.[61] Lobbying by a local group that had moved beyond entrenched highlander/lowlander disputes to address conditions throughout a watershed prompted the Manitoba government to do what it had long resisted – pass legislation to enable environmental management on a watershed basis.[62] The bridging of a long-standing fissure in public opinion made it easier for provincial administrators to establish the legislative infrastructure necessary for environmental management according to the watershed idea. And the government demonstrated an ongoing willingness to accommodate the concerns of interested local residents with the 1959 passage of The Watershed Conservation Districts

Act, which addressed what were perceived to be the deficiencies of the 1958 legislation.

Despite all that, no watershed-based districts were formed in the 1960s. For officials moved to action by the efforts of the Riding Mountain-Whitemud River Watershed Committee, it must have been especially frustrating that no district was formed in this area. The trouble was due in part to processes playing out in Manitoba's wet prairie as in rural areas across Canada and the United States. Changes in farm and transportation technologies as well as the globalization of agricultural markets had altered the economics of production. Many farm families believed that brighter futures lay in towns or cities and left their land in search of greater prosperity elsewhere.[63] The population of the Whitemud watershed dropped from 25,460 in 1941 to 21,571 in 1971 (15 percent); the rural component declined from 23,168 to 15,662 (32 percent) over the same period.[64] Those who remained in agriculture often enlarged their operations, buying up the land that their neighbours were selling and investing in the new farm machinery that made it possible to work enormous acreages.

In this changed human landscape, rural municipal finance presented particular challenges. Between 1951 and 1962, per capita tax impositions by rural municipalities increased by 59.2 percent. Taxes went up because municipal spending increased (by some 68.7 percent). Rural municipalities spent a greater proportion of their budgets on public works than did villages, towns, suburbs, and cities. Indeed, their public works costs accounted for 28.58 percent of their expenditures, as compared with a maximum of 15.91 percent for the others.[65] Watershed management could operate as a means to leverage government funding for local projects, but some of the associated costs would have to be borne locally. Given the substantial proportion of municipal revenues already expended on public works projects, perhaps it is not surprising that rural municipal governments were reluctant to take on a potentially expensive new endeavour. Since rural municipalities relied heavily on revenue derived from real property through personal property taxes and special assessments, increased municipal expenditures would likely have translated into increased tax rates. Also, the possibility of school district consolidation driven by population changes led to particular concern with school taxes during the years in which advocates were promoting watershed management.[66] Even members of the Riding Mountain-Whitemud River Watershed Committee, the very group that had worked toward enabling legislation, were reluctant to create conditions that might lead to increased taxes. Indeed, committee members lobbied the government to lower the maximum mill rate that could be imposed on the watershed, from ten to five, from five to three, and then even down to two mills.[67]

Concerns about increased financial burdens were compounded by fears about diminished municipal autonomy as rural communities struggled with rapid demographic changes and cast a leery eye on arrangements that might transfer authority from locals to provincial officials. Some people feared that the creation of another administrative entity would diminish both the power of their municipal leadership and the sense of community they shared with their fellow municipal residents.[68] Another problem was with the relatively short tenure of municipal councils. The benefits of the watershed approach would become apparent over the long term, yet the careers of many municipal politicians were relatively short. Municipal councillors were being asked to enact a policy that might pay big dividends in the long term but that would surely cause administrative hassles in the immediate future. There was a troublesome disconnect between the political boundaries and time frames already in place and those associated with watershed management. All of these factors, from the financial to the logistical, added up to a significant barrier to the formation of a watershed-based conservation district.

Financial considerations not only helped to forestall the establishment of watershed authorities but also interfered with the continued operation of the existing system of drainage maintenance districts. This system had been in operation since the 1895 Land Drainage Act and persisted through the passage of the 1958 Watershed and Soil Conservation Authorities Act and the 1959 Watershed Conservation Districts Act. But in the mid-1960s, spurred by the recommendations of the Manitoba Royal Commission on Local Government Organization and Finance, drainage maintenance districts were dissolved. In the hope of achieving more efficient surface water management, it was proposed that responsibility for drains of a certain magnitude would fall on the provincial government and that municipalities would look after smaller waterways.[69]

With drainage maintenance districts dissolved and no watershed and soil conservation authorities yet created, there was no mechanism for the coordinated administration of drains that spanned municipal boundaries. Donald Pisani has identified "the proliferation of special districts" such as drainage districts as "the untold story" of natural resources administration in the twentieth century.[70] The Manitoba government sought to counter the international trend only briefly, for the challenge of managing drainage without a dedicated administrative infrastructure soon became apparent.[71] The 1967 creation of the Manitoba Water Commission, an independent body formed to investigate various problems related to water management, reflected renewed appreciation for the advantages of government agencies geared to environmental administration.[72]

Any new perspective the Manitoba Water Commission might have offered, however, was quickly submerged by the prevailing administrative confusion surrounding drainage. From Manitoba's earliest days, the Department of Public Works had responsibility for drainage. But its focus on physical infrastructure was not consistent with a more holistic conception of drainage as one aspect of a larger surface water and land management issue. Responsibility for drainage was bandied between the Department of Public Works, the Department of Agriculture, the Department of Highways, and the Department of Mines and Natural Resources as the province groped for an appropriate bureaucratic structure.[73] In 1970, the situation changed again as the province passed The Resource Conservation Districts Act. This legislation was distinguished by a key characteristic: it enabled the formation of districts defined along municipal lines rather than watershed boundaries.[74] This was an attempt to woo reluctant municipalities by further accommodating their fears about the loss of local autonomy and the destruction of municipal community. It had the immediate effect, however, of further complicating what was already a confused administrative situation. If the environmental conditions relevant to Manitoba drainage were complex, by the mid-twentieth century the human governmental context was equally so.

In 1972, the long-anticipated Whitemud Watershed Conservation District was finally formed under The Watershed Conservation Districts Act of 1959. District boundaries approximated a "natural catchment area" that was to be governed by "a single authority" concerned with establishing "a unified, co-ordinated approach to the development of schemes and systems for managing and controlling the water resources of a district and in conjunction with this, the land, forest, wildlife, and recreational resources" of the area.[75] The establishment of the Whitemud Watershed Conservation District was seen by Manitoba officials as "a major step forward in this Province towards wise management of our soil and water resources." The hope was that it might prove "a turning point in the preservation of our precious natural resources" by setting a precedent that other areas would emulate.[76]

The establishment of this district was also an attempt to take advantage of new opportunities to secure federal financial assistance. Over the second half of the twentieth century, the Canadian federal government became increasingly concerned with regional disparities and established programs to promote development in disadvantaged areas. Created in 1961, the Agricultural Rehabilitation and Development Act (ARDA; later retitled the Agricultural and Rural Development Act, same acronym) provided for cooperation by the federal and provincial governments on cost-shared undertakings, including initiatives aimed at more effective land and water management.[77] The improvement of the Norquay Floodway, a substantial drain running through the Red River valley to the west of the Red River, was the first ARDA construction project approved by both levels of government.[78] By January 1968, over $10.6 million in ARDA funds had been spent on or committed to improving drainage.[79] The increasingly significant idea of regional development provided a new logic for federal involvement in what had long been understood, in Manitoba at least, as a provincial sphere of responsibility.

Development work across Canada was restructured and expanded with the 1969 creation of the Department of Regional Economic Expansion, which was to provide "a more integrated and coordinated set of policies for regional economic development."[80] The new department absorbed numerous existing programs and assumed responsibilities as defined by multiple pieces of legislation, including the Agriculture and Rural Development Act. By the early 1970s, negotiations between Winnipeg and Ottawa were under way for ARDA III, the third federal-provincial agreement governing programs administered under the 1961 legislation. This new agreement maintained the existing emphasis on improved land and water management but changed how this goal was to be pursued. "Unlike former ARDA Agreements," a joint publication by the provincial Department of Agriculture and the federal Department of Regional Economic Expansion explained, "the current thrust is a multi-dimensional effort limited to selected target areas."[81] ARDA III provided for a Comprehensive Soil and Water Conservation Program, with a projected budget of $11,256,450 over five years.[82] The costs would be split equally between the provincial and federal governments.[83] Among the program's activities was the promotion of watershed-based approaches to environmental challenges within select areas of the province, including the Whitemud River watershed.[84] Projects undertaken could be administered by either the province or the local conservation district, meaning that it was not absolutely necessary to form a district to secure some of the available funds. Nevertheless, the program represented a significant new incentive to district creation. The formation of the Whitemud Conservation District in the early 1970s reflected not only the efforts of local activists and provincial officials but also the impacts of federal regional development programs.

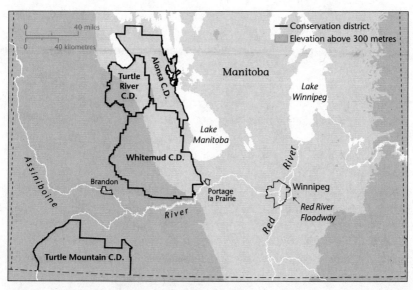

FIGURE 20    Early conservation districts in Manitoba
*Source:* Adapted from Conservation Districts in Manitoba, map, Manitoba Water
Stewardship, 2004.

The formation of conservation districts in the Turtle Mountain and
Turtle River Watersheds followed in 1973 and 1975 respectively.[85] Both were
situated in areas vulnerable to erosion.[86] And both were among the areas
targeted by ARDA III. The latter district was established under Manitoba's Watershed Conservation Districts Act (1959) and targeted in ARDA's
Water Conservation Sub-Program along with the Whitemud Conservation District. The former was established under Manitoba's Resource
Conservation Districts Act (1970) and included in ARDA's Soil
Conservation Sub-Program, which, along with the Water Conservation
Sub-Program and the Alternative Land Use Program, made up the Comprehensive Soil and Water Conservation Program.[87] Yet, despite these two
successes, there was still no general rush to form districts.

In the mid-1970s, the Manitoba Water Commission was instructed to
review agricultural drainage in the province. After assessing physical infrastructure, governing legislation, and public opinion, commissioners endorsed a return to more extensive special district government. They argued
that the 1970 Resource Conservation Districts Act should "be used as a
mechanism to ensure the adequate maintenance, upgrading and expansion
of agricultural drainage systems."[88] Some amendments were proposed,
but the existing legislation was seen as satisfactory in its major aspects.

Commissioners emphasized that the provincial government should actively promote the formation of conservation districts.[89] The Alonsa Conservation District was established in 1978. But this one success aside, and despite the endorsement of the Manitoba Water Commission, it was clear by the late 1970s that district formation would be a slow process. Indeed, until the mid-1990s, there were only six districts in the province.[90]

The varied challenges to the expansion of the conservation district system were exacerbated by the history of contention over surface water management in Manitoba.[91] Even as highlander interest in watershed management reshaped long-standing patterns of dispute, and even as federal regional development programs relaxed the financial constraints on drainage, dissatisfaction and disappointment persisted. The stasis that fell over the administration of Manitoba drainage in the third quarter of the twentieth century was not specifically an outcome of the various settlement processes, jurisdictional dynamics, or conceptual debates that characterized drainage in Manitoba. Nor was it driven by highlander/lowlander rivalries. Rather, it was a product of the legacy of discontent and distrust that emerged from all of these various issues. The history of contention over drainage made it difficult to generate public and administrative support for any sort of new initiative, even one that promised to address the problems of the past.

William R. Newton, in his role as director of the Manitoba Water Resources Branch, provided a vivid description of the situation in an April 1973 presentation to a meeting of forest engineers. Newton explained how every year, on visiting afflicted areas, he had

> very little difficulty in finding people who will tell me exactly what the problem is – what the solutions are – people who will be quite specific about who is to blame. I am told that the upstream municipalities are draining off their waters too fast. I hear that neighbouring municipalities are building too many roads, or are building up roads to act as dykes to hold the water on the lands of another ... I am told by one group that the Water Resources Branch are building too many drains or that the drains they are building are too big or too small; by another that drainage is being neglected. As far as who is responsible it always seems to be someone else and it always seems that the solution rests with the Province or with the engineer.[92]

Newton recognized the benefits of watershed conservation districts and advocated the immediate adoption of watershed management. Indeed, most of the text of his talk was a description of the advantages of such an approach. But his discussion of the current situation recognized the importance not only of the specific "comments and criticisms" of the public but also of how all expressions of dissatisfaction became part of what he called a "dialogue" – a prevailing tone in much communication over drainage that reflected the dissatisfaction and distrust felt by many Manitobans.[93] If an earlier generation had taken a far too optimistic view, assuming that surface water flooding could be permanently solved, mid-twentieth-century Manitobans were profoundly jaded, finding human incompetence where they might have seen environmental complexities and administrative challenges.

Manitoba farmer Allan Chambers confirmed in 1984 that "not many of us are prepared to discuss drainage calmly and rationally." He went on to explain that, in afflicted regions, drainage could be added "to religion and politics as topics to avoid in polite company."[94] Although people were curious about new approaches to persistent problems, their frustration at what were perceived to be past government failures contributed to pessimism toward new water management strategies. Although such sentiments were certainly more entrenched among lowlanders, confusion over policy changes quickly opened what one Manitoba government employee termed a "credibility gap" between highland advocates and provincial administrators.[95] Highlanders and lowlanders were united in their distrust of provincial initiatives. The public hearings conducted by the Manitoba Water Commission made it clear that problems existed "between all possible combinations of individuals, municipalities and the Provincial Government."[96] After decades of protracted dispute, surface water management in Manitoba was as much the management of contention as the management of water. The negative tone that dominated public discourse over drainage for more than seventy years complicated the deployment of new ideas. Many people formed opinions based on past frustrations as much as on new proposals. Ultimately, by the late twentieth century, public dissatisfaction had become an environmental problem in its own right. The human history of drainage in the region had itself become a management challenge every bit as daunting as surface water.

Amid this troublesome human history, apparently as part of an effort to appeal to reluctant Manitobans rather than as a consciously chosen policy direction, attempts to address Manitoba's surface water problems moved away from the idea of watershed management. When the 1959 Watershed Conservation Districts Act and the 1970 Resource Conservation Districts Act were consolidated in the 1976 Conservation Districts Act, districts could be created according to either watershed or municipal boundaries or a mix of the two. After the mid-1990s, district formation increased in pace. By the early twenty-first century, there were eighteen districts in operation, covering approximately 85 percent of municipal Manitoba.[97] The vast majority of these districts, however, defined their areas of concern according to municipal rather than watershed boundaries.[98] A recent report by the International Institute for Sustainable Development (IISD) argued that, in the absence of widespread adoption of the principle of watershed management, Manitoba's conservation district system "merely created an additional layer of administration" without dealing progressively or even effectively "with the challenge of surface water management."[99]

## CONCLUSION

By the middle decades of the twentieth century, much was changing in the wet prairie landscape. The dynamic character of the provincial hydroclimate had become clear, prompting Manitobans to move beyond the ideal of permanence to consider how the drainage infrastructure could be adapted to a variable environment. The decision to tie drainage funding more closely to environmental conditions was important. With available funds rising and falling in relation to the amount of water in the surrounding ecosystem, the drainage infrastructure was less likely to be compromised by inadequate resources in wet periods. It took a while longer for watershed management to gain favour. Eventually, however, new concerns over surface water erosion led to a reconceptualization, though halting and tentative, of the drained landscape.

Even when acknowledged as an admirable goal in the abstract, watershed management was challenging to enact in a landscape already crisscrossed with drains that had been located in relation to other considerations. But it is important also to recognize that Manitobans were caught in a trap made up as much of failed efforts as of physical infrastructure. Many remained unconvinced that administrative adjustments would resolve decades of disappointment and dissatisfaction, even with the benefit of the federal funds that were available once drainage became a regional development activity. By the mid-twentieth century, discontent was as persistent as surface water in Manitoba's wet prairie. Indeed, it generated a momentum of its own that amounted to a conceptual rigidity most evident in reluctance to adopt the watershed approach.

While accrued negativity hung heavily over all discussions of drainage in the second half of the twentieth century, rural land drainage was changing.

To farmers, surface water was as important as ever. But the province was increasingly urbanized, and city water problems were of another order.[100] A key concern after the Red River flood of 1950 was the question of catastrophic flooding in urban areas. The Red River Floodway can be seen as a measure of environmental maladjustment, marking the sort of dramatic environmental change necessary to protect established patterns of human settlement. But the construction of the floodway in the 1960s also reflected the new realities of a province dominated by a growing urban centre and the commitment of the provincial government to respond to the needs of city voters. In rural areas, fewer agriculturalists were working larger farms. Overall, the provincial economy was increasingly diversified. For all these reasons, the portion of the population profoundly interested in drainage was diminishing. Furthermore, when reconceived as a regional development activity, drainage was absorbed into a larger federal bureaucracy that also dealt with a wide array of problems across the country. Federal money and authority came to bear on surface water management, an area of activity that once had been almost exclusively the responsibility of the province, and changed the intellectual environment in which water management decisions were made. As a result of all of these factors, the context for drainage in Manitoba had shifted dramatically. Yet at least one thing remained the same: by the end of this period as at the beginning, the principle of watershed management did not significantly bear on surface water administration in the province.

In the early years of the twenty-first century, surface water management remains an important activity of the provincial government. In agricultural Manitoba, complaints about inadequate drainage seem to be as perennial as the arrival of spring. An April 2008 report on the licensing and enforcement practices of Manitoba Water Stewardship by the Manitoba ombudsman dealt primarily with continued contention over drainage. The document makes clear that, despite further changes in administrative structure and management procedure over the past few decades, dissatisfaction remains widespread.[101] After a devastating flood along the Red River in 1997, the International Joint Commission made recommendations that indicated the need to expand the floodway.[102] Over a decade later, construction was still under way. Even considering economic diversification and continuing urbanization and changed government programs, providing security and stability for both rural and urban Manitobans is an ongoing challenge in a landscape that remains profoundly dynamic and persistently wet.

CONCLUSION

# Chequer Board Squares in a
# Dynamic Landscape

Soon after the 1870 creation of the province of Manitoba, the head of
the new provincial government instructed Legislative Clerk F.E.
Molyneux St. John to study local factors that might affect agricultural
settlement. St. John's report, submitted to Lieutenant Governor Adams
Archibald a few months later, emphasized "the number of muskegs or
swamps which are found in several parts of the Province." Although St.
John recognized that they might prove to be valuable as sources of hay
and water, he understood that wet areas could interfere with cropping. "It
is therefore not unreasonable to assume," he concluded, that the "land
which a settler should be entitled to take up must in some measure depend
on the nature of the country." In his view, the shape and size of the parcels
of land made available to newcomers should reflect their environmental
characteristics.[1]

St. John's report was an important early recognition of the need to accommodate water patterns in the wet prairie. The report did not, however,
succeed in affecting dominion land policy. Keen to settle the North
American northwest as rapidly as possible, and influenced by policy developed elsewhere, Canadian officials charged with administering the
prairie landscape embraced a grid-based settlement plan. Indeed, even
experts who recommended deviating from the standard survey in some
particularly challenging landscapes thought that southern Manitoba was
entirely amenable to such treatment. For example, when confronted with
the "tossed sea of Mountains" in British Columbia, surveyor Lindsay
Russell recommended that "the system of division of any habitable lands

into agricultural or other holdings must be a natural one." In such dramatic terrain, there was no way that any "artificial uniformity" could be imposed "without detriment to the future dwellers in the land." In contrast, Russell was convinced that, in "the level country east of the Rocky Mountains," the division of land into "chequer board squares" was "the best possible" arrangement for the area.[2]

Although it took longer than Manitoba boosters had hoped, home-steaders eventually showed up in large numbers and chose their squares. Many set about establishing farms before taking a hard look at the country in which they found themselves. Early choices made in haste or for ill-considered reasons could have long-term consequences that became apparent only over time as the dynamic character of the wet prairie revealed itself in how it veered from a landscape amenable to agriculture to the extremes of water overabundance. In Manitoba, the private tragedy of the homesteader who laboured long and hard to plant a crop in a location vulnerable to intermittent inundation paralleled the decision of the federal government to run the nationally significant transcontinental railway through a flood-stricken route despite available alternatives. The problem was not primarily arrogance or optimism, though both certainly figured. At root, the trouble was a lack of understanding. Only over time did it become apparent that the lands of Manitoba were not as susceptible to "artificial uniformity" as had been thought. When the surface water problem presented itself, it did so in dramatic ways, with washouts of essential roads and drownings of field crops. For many farmers in Manitoba's wet prairie, the significance of variations in water levels from year to year and season to season came to loom as large as the Rocky Mountains.

Both governments, the provincial and the federal, recognized the importance of drainage to the future of the province. Providing necessary assistance of this sort was consistent with the liberal principle of facilitating capital accumulation, which at this time and place amounted to supporting agriculturalists. The need for drainage became intertwined with pressing jurisdictional disputes between the two governments, and politics and the environment bore on each other as difficulties with counting and mapping the wet prairie complicated any resolution. Ultimately, the province was left to figure out how to help settlers who had unknowingly settled in areas vulnerable to flooding. For lowland farmers, geographical location became a common experience layered atop differences in language and culture. Insofar as large-scale efforts to address surface water problems required government management and financing, the lowland community of interest also shared a parallel orientation to the expanding provincial state.

Although inappropriate practices such as patronage did not help matters, drainage proved to be challenging largely because human efforts at environmental change were embedded within long-standing environmental patterns. The season-to-season, year-to-year, decade-to-decade changes in surface water abundance provided the unstable context for efforts to transform the wet prairie. The success or failure of drainage efforts could have as much to do with larger environmental patterns as with the work itself. But Manitobans did not always recognize this. Many were quick to lay blame for problems that could not in truth be attributed to human negligence. Over time, discontent and frustration became additional factors uniting drained areas.

Faced with persistent flooding, lowlanders eventually began to connect their misfortunes to the actions of those farmers situated at higher elevations, claiming that increased runoff from highland areas was impeding lowland drainage. Conflict between highlanders and lowlanders revolved around the contentious matter of how to define the liberal ideas of property and progress within Manitoba's wet prairie. Highlanders held to definitions that seemed to them to be consistent with the township survey, arguing for their right to do as they wished on their lands. Lowlanders argued for what they believed to be more environmentally attuned conceptions under which highland farmers would be obliged to consider the lowland consequences of their actions. These divergent colloquial liberalisms illustrate the effect of geography on how non-experts interpreted and applied liberal ideas. Contention between these two perspectives makes clear that the process of learning about the local environment was not merely a question of spending time in the region. People perceived the environment in different ways and became convinced of the validity of different management strategies based on their own interests. As drainage continued apace, contention only increased.

In the late 1910s and early 1920s, the province engaged drainage experts and commissioned focused studies in a bid to resolve the increasingly problematic matter of surface water flooding. Although this process played into a change in the nature of provincial administration, with an increased reliance on professional expertise, it failed to produce a broadly accepted solution to the drainage problem. Eventually, by the mid-twentieth century, Manitobans came to favour the province taking a larger role in drainage funding, based on an increased appreciation of the inevitability of environmental variability in the wet prairie as well as a greater willingness to use government authority to protect those who were vulnerable. But despite this, and despite evidence of progressive environmental thought in relation to international flooding along the Roseau River and waterfowl preservation by Ducks Unlimited, factors such as institutional inertia, continued agricultural intensification, and the by then substantial investment of money and effort in the existing drainage system meant that drainage practices remained relatively unchanged through the 1940s and 1950s.

Ultimately, in prompting change in Manitoba's environmental administration, newly apparent environmental problems were at least as important as gradually acquired environmental knowledge and broad changes in social context. Soil erosion by surface water, a factor in some areas of Manitoba by the mid-twentieth century, proved to be an important catalyst to the reconceptualization of land management in the wet prairie. Many afflicted highland farmers began to consider whether a new type of environmental management, one oriented to the watershed, might provide a more effective means of combatting erosion. The emerging highland interest in watershed management accorded with what lowlanders had long sought: the creation of a management entity that would act in what was now seen as the common interest of all who were connected by water flows.

But even as they came to share an interest in watershed management, highlanders and lowlanders were also united by their distrust of each other and government initiatives. Combined with broad changes in the agricultural landscape, this accrued pessimism represented a significant barrier to the adoption of watershed management. As Manitoba provincial government employee and engineer F.E. Umphrey made clear, drainage had been judged "unsatisfactory and insufficient ever since it was introduced."[3] Generated largely by the failed efforts of the past and compounded by the administrative challenges to such a dramatic shift in surface water management, general dissatisfaction and distrust prevailed. Such widespread negativity in relation to Manitoba drainage was more than an unfortunate outcome of what many perceived as largely unsuccessful efforts at environmental change. Insofar as the adoption of watershed management required the cooperation of Manitobans at large, this pervasive discontent amounted to an environmental management problem in its own right. Government attempts to appeal to reluctant and dissatisfied Manitobans included the development of legislative mechanisms to enable the creation of districts that reflected meaningful human boundaries, such as those between rural municipalities. Whether or not this strategy succeeded in easing fears about changes in rural life, it was certainly a deviation from what had seemed like a trajectory toward watershed-based administration.

Provincial efforts to address the consequences of persistent negativity, then, ultimately contributed to steering the province away from watershed management.

Within this environmental history of Manitoba's wet prairie are some important connections to broader questions. The introduction invoked three key scholarly landmarks: the state, liberalism, and the challenges of surface water management in a human world of boundaries and jurisdictions. At the broadest scale, this study offers a basic insight relevant to scholarship on the state and liberalism: the local environment, in this case the wet prairie, affects how government authority is deployed and how ideological principles are enacted. In a manner consistent with liberal principles, the Manitoba provincial government early on extended itself to take on the pressing task of drainage. Municipalities and eventually drainage districts were created, becoming local mechanisms through which surface water problems were addressed. Interested Manitoba residents kept one eye on their government and the other on their land, evaluating each in the context of the other. Manitobans also watched each other and developed variants on the pervasive liberal ideology that reflected their geographical position in the province. Both thinking about liberalism and the operation of the state reveal the impress of Manitoba's geography.

Furthermore, Manitoba's troublesome landscape provides a particularly useful vantage point from which to consider the relations between state and environment. The mismatch between liberalism (with its emphasis on individualism and private property) and the wet prairie environment (with its water flows that continue in disregard of property boundaries) prompted state officials, drainage experts, and Manitoba residents to seek to redesign the landscape and reimagine liberalism in an effort to bring the two into better alignment. Because the success of agriculture depended to some degree on this reconciliation, and because agriculture maintained an important role in the provincial economy, the misalignment could not simply be ignored or avoided. Ultimately, the mismatch between liberalism and landscape, as well as attempts to resolve it, revealed both the character of the province's environment and the character of liberalism as deployed in the province. For those who study topics beyond agriculture and the environment, the wet prairie landscape in itself might seem to be of little interest. But liberalism, as a cultural ideology also significant in other times and places, might be of wider concern. This study suggests that scholars interested in political culture might consider examining the management of and debates surrounding signal landscapes, those that seem to be particularly likely to cast into relief broadly significant cultural ideologies.

From this perspective, environmental history might be a method as well as a subfield.

*Wet Prairie* also documents the persistent efforts of various actors, from governments to settlers, to grapple with human constructs such as borders or jurisdictions on the one hand and environmental realities such as surface water on the other. In Manitoba, there were a number of significant mismatches between human and environmental geographies, from a grid-based settlement system in a wet and variable landscape through an international border that fractured drainage patterns to drainage districts and, later, conservation districts that encompassed only part of the relevant watersheds. Many Manitobans were profoundly disappointed with the management of at least some of these situations. *Wet Prairie* makes clear how, for dissatisfied Manitobans, these mismatches added up over time to a persistent negativity and distrust directed toward the provincial government. Although not necessarily directly or obviously connected to each other, these mismatches all contributed to the creation of the unfavourable context in which new provincial government initiatives were evaluated. This situation rendered only more complicated efforts to bring human and environmental geographies into better alignment.

In Manitoba, the prairie is wet. This simple observation is significant to those interested in the history of the province, the prairie region, or the North American environment. The Canadian prairie is often described as an area of water scarcity, but much of southern Manitoba just does not make sense in these terms. The development of the Manitoba provincial state and the experiences of many of its residents reflect the influence of a different set of environmental circumstances. A greater average rate of precipitation, relatively impermeable soil in important agricultural areas, and provincial topography that exacerbates surface water problems in some locations: these factors combine to ensure that many Manitoba farmers have worried at least as much about flood as drought. Considering also the intermittent threat of catastrophic inundation along the Red and Assiniboine Rivers, it is evident that much of southern Manitoba has been influenced more by water excess than by water scarcity.

This observation is also significant to contemporary Manitoba residents and administrators grappling with what they see as surface water problems, with the important additional assertion that the wet prairie is a highly variable landscape. In this region, environmental change is endemic. Local surface water problems are unpredictable in manifestation and enduring in complexity. Importantly, this regional reality is not necessarily immediately evident to a farmer on the land; rather, it becomes more readily apparent in relation to a larger swath of the province considered over a longer period of time. Manitobans have recognized the need for improved public understanding of environmental conditions. At a 1958 seminar put on by the provincial government for those involved in municipal admin- istration, the watershed approach to surface water management was the topic of discussion. E.H. Poyser of the Department of Agriculture offered an introduction to the concept of the watershed and an overview of how this idea might affect environmental management.[4] In the discussion that followed his presentation, questions from the audience prompted Poyser to reflect on the difficulties of achieving public support for watershed management. "Perhaps," he mused, "our whole educational programme in conservation in Manitoba has been inadequate." Although farmers knew their own land intimately, many lacked any real understanding of the broad environmental challenges of surface water management across the wet prairie. Poyser's comment was well received and echoed in subsequent contributions to the discussion. "You cannot," one municipal representa- tive commented, expect a widespread or an immediate embrace of water- shed management "without the people knowing all the facts."[5] By the middle of the twentieth century at least, the need for improved public understanding of the wet prairie landscape was, to those who reflected on such matters, as blatantly apparent as continued problems with surface water management.

By the time Poyser was meeting with municipal officials in the late 1950s, Manitobans had been waiting a long time for solutions to what they perceived to be environmental problems. In this situation, frustration is understandable. But the difficulty is that environmental conditions in the wet prairie are not problems that can be solved. Beyond the simple reality of surface water, the most basic fact about the wet prairie is change over time. Recurring difficulties with drainage are not due to administrative incompetence or inappropriate land management systems, though they have certainly exacerbated matters. Rather, intermittent inundation is often simply evidence of the character of the province's environment. Manitoba needs a broader public understanding of the role of environmental vari- ability in local surface water management. In the wet prairie, where surface water is often abundant, it is such understanding that has proven to be in short supply.[6] There are particular challenges to cultivating understand- ing in a place where vulnerability to flooding is unevenly distributed and where alternating periods of exacerbated or diminished flood risk occur in time frames that span generations.[7] Nevertheless, this might be the most effective way to counter the negativity that developed as a consequence of what many have seen as the province's failed drainage efforts and that ultimately interfered with the adoption of new surface water management strategies. Manitoba is wet prairie, and the wet prairie is a highly variable landscape. In a landscape so changeable, improved public understanding of these local environmental realities might help to provide a more solid foundation for the future.

# Appendices

Table A.1

**Annual capital expenditures on drainage, approximate to nearest thousand, 1896-1932**

| Year | Expenditure($) | Year | Expenditure($) | Year | Expenditure($) |
|------|------|------|------|------|------|
| 1896 | 42,000 | 1909 | 142,000 | 1921 | 246,000 |
| 1897 | 47,000 | 1910 | 299,000 | 1922 | 141,000 |
| 1898 | 56,000 | 1911 | 249,000 | 1923 | 214,000 |
| 1899 | 132,000 | 1912 | 178,000 | 1924 | 306,000 |
| 1900 | 72,000 | 1913 | 389,000 | 1925 | 225,000 |
| 1901 | 85,000 | 1914 | 352,000 | 1926 | 228,000 |
| 1902 | 89,000 | 1915 | 288,000 | 1927 | 249,000 |
| 1903 | 357,000 | 1916 | 42,000 | 1928 | 4,000 |
| 1904 | 230,000 | 1917 | 75,000 | 1929 | 416,000 |
| 1905 | 339,000 | 1918 | 19,000 | 1930 | 179,000 |
| 1906 | 315,000 | 1919 | 4,000 | 1931 | 73,000 |
| 1907 | 344,000 | 1920 | 128,000 | 1932 | 41,000 |
| 1908 | 120,000 | | | | |

*Note:* The approximate total of all expenditures in this period was $7 million.
*Source:* All figures are from Archives of Manitoba, GR 1609, G 8046, file Union of Municipal Drainage Maintenance Districts, text of an address delivered by F.E. Umphrey at the annual meeting of the Union of Municipal Drainage Districts, 27 November 1945.

## APPENDIX B

TABLE A.2

**The status of drainage districts in 1934**

| District number | Lands benefited (acres) | Drains constructed (miles) |
|---|---|---|
| 1 | 62,760 | 70.00 |
| 2 | 449,591 | 1,181.36 |
| 3 | 36,364 | 120.75 |
| 4 | 80,508 | 160.50 |
| 5 | 130,206 | 92.20 |
| 6 | 21,270 | 92.49 |
| 7 | 8,400 | 6.00 |
| 8 | 393,981 | 461.67 |
| 9 | 140,059 | 173.50 |
| 10 | 43,610 | 68.00 |
| 11 | 70,094 | 62.50 |
| 12 | 132,776 | 276.59 |
| 13 | 7,232 | 6.20 |
| 14 | 67,088 | 109.75 |
| 15 | 32,642 | 30.00 |
| 16 | 64,045 | 151.20 |
| 17 | 34,006 | 20.60 |
| 18 | 39,192 | 34.00 |
| 19 | 162,898 | 299.34 |
| 20 | 107,414 | 177.05 |
| 21 | – | – |
| 22 | 9,390 | 10.75 |
| 23 | 9,828 | 29.45 |
| 24 | 4,800 | 8.75 |
| *Total* | 2,108,154 | 3,642.65 |

*Note:* Drainage District No. 21 was proclaimed, but no construction was undertaken.
*Source:* William P. Elliot, "Artificial Land Drainage in Manitoba: History – Administration – Law" (MRM thesis, University of Manitoba, 1977). Elliot is drawing on the Government of Manitoba, Department of Public Works, Annual Report for 1934.

## Dates of Relevant Treaties, Reports, Events, and Legislation

1867 Canada is created through the British North America Act.

1870 Manitoba is created through the Manitoba Act.

1872 Canada passes the Dominion Lands Act.

1880 Manitoba passes The Drainage Act.

1885 Canada passes the Swamp Lands Act.

1894 Canada passes the Northwest Irrigation Act.

1895 Manitoba passes The Land Drainage Act.

1909 The United States and Great Britain (on behalf of Canada) sign the Boundary Waters Treaty.

1916 The United States and Great Britain (on behalf of Canada) sign the Migratory Bird Treaty.

1919 The Manitoba Drainage Commission is established, chaired by J.G. Sullivan.

1921 The *Report of the Manitoba Drainage Commission* (also termed the Sullivan report) is submitted.

1930 Control over natural resources is transferred to the prairie provinces.

1935 The Land Drainage Arrangement Commission is established, chaired by John N. Finlayson.

1936 The *Report of the Land Drainage Arrangement Commission Respecting Municipalities Containing Land Subject to Review under "The Land Drainage Act"* (also termed the Finlayson report) is submitted.

1937 Ducks Unlimited and Ducks Unlimited (Canada) are incorporated.

1947 M.A. Lyons begins an investigation into Manitoba drainage.

1950 The *Report and Recommendations on "Foreign Water" and Maintenance Problems in Drainage Maintenance Districts Constituted under the Land Drainage Arrangement Act, 1935, Province of Manitoba* (also termed the Lyons report) is submitted.

1950 There is catastrophic flooding along the Red River.

1958 Manitoba passes The Watershed and Soil Conservation Authorities Act.

1959 Manitoba passes The Watershed Conservation Districts Act.

1961 Canada passes the Agricultural Rehabilitation and Development Act (later retitled the Agricultural and Rural Development Act).

1967 The Manitoba Water Commission is created.

1970 Manitoba passes The Resource Conservation Districts Act.

1976 Manitoba passes The Conservation Districts Act.

# Notes

## Foreword

1 Aldo Leopold, *A Sand County Almanac and Sketches Here and There* (New York: Oxford University Press, 1949). Paperback edition 1968; Daniel Berthold-Bond, "The Ethics of 'Place': Reflections on Bioregionalism," *Environmental Ethics* 22, 1 (2000): 5-24.

2 Leopold, *Sand County Almanac*, 95 and 96.

3 Ibid., 158, 160 and 158.

4 Jennifer M. Shay, "Vegetation Dynamics in the Delta Marsh, Manitoba," in *The Prairie? Past, Present and Future: Proceedings of the Ninth North American Prairie Conference*, ed. Gary K. Clambey and Richard H. Pemble (Moorhead, MN: Tri-College University Centre for Environmental Studies, 1986), 65-70.

5 Quotes in this paragraph from Leopold, *Sand County Almanac*, 159 and 162. Delta Marsh has not yet been put to the plow, but engineering works initiated in 1961 have stabilized water levels in the lake and produced a steady expansion of emergent vegetation. It was designated in 1982 as a Wetland of International Significance under the Ramsar Convention. For an account of some of the consequences of the 1961 works see Jennifer M. Shay, Petronella M.J. Geus, and Margaret R.M. Kapinga, "Changes in Shoreline Vegetation Over a 50-year Period in the Delta Marsh, Manitoba in Response to Water Levels," *Wetlands* 19, 2 (1999): 413-25.

6 Daniel Berthold, "Aldo Leopold: In Search of a Poetic Science," *Human Ecology Review* 11, 3 (2004): 205-14, quote on 207.

7 Quotes in this paragraph from Leopold, *Sand County Almanac*, 160.

8 Ian McKay, "The Liberal Order Framework: A Prospectus for a Reconnaissance of Canadian History," *Canadian Historical Review* 81, 4 (December 2000): 617-45.

9 A.B. McKillop has suggested that McKay's article engaged "Canadian historians to a degree not witnessed since Careless's 'limited identities' article inspired a generation of fledgling social historians in the seventies." A.B. McKillop, "From the Editor's Desk," *The Underhill Review* (Fall 2009), Online. For limited identities see J.M.S. Careless, "'Limited Identities'

in Canada," *Canadian Historical Review* 50, 1 (March 1969): 1-10. McKillop also situates the liberal order framework alongside W.L. Morton's critique of the Laurentian Thesis written in 1946, "Clio in Canada: The Interpretation of Canadian History," originally published in *University of Toronto Quarterly*, 15 (April 1946): 227-34, republished in *Contexts of Canada's Past: Selected Essays of W.L. Morton*, ed. A.B. McKillop (Toronto: Macmillan, 1980), 103-12 as one of three interpretive turning points in Canadian historical scholarship. McKay's work has inspired both acolytes and critics. In a commentary on McKay's work, Andrew Smith has suggested, without systematic investigation, that half the recent dissertations in Canadian history invoke McKay's framework (http://andrewdsmith.wordpress.com/2010/07/14/mckays-liberal-order-framework-promise-and-pitfalls/). For critique and analysis see Michel Ducharme and Jean-François Constant, eds., *Liberalism and Hegemony: Debating the Canadian Liberal Revolution* (Toronto: University of Toronto Press, 2009) and essays by Janet Ajzenstadt, Jean-Marie Fecteau, Martin Paquet, and Nancy Christie in *The Underhill Review* (2009).

10  McKay, "The Liberal Order Framework," 623.

11  Ibid., 620-21.

12  Jean-Marie Fecteau, "Towards a Theory of Possible History? Ian McKay's Idea of a 'Liberal Order,'" *The Underhill Review* (Fall 2009).

13  Bryan D. Palmer, "Radical Reasoning," *The Underhill Review* (Fall 2009).

14  Ian McKay, "Canada as a Long Liberal Revolution: On Writing the History of Actually Existing Canadian Liberalisms, 1840s-1940s," in *Liberalism and Hegemony*, ed. Ducharme and Constant, 347-452, quote from 349.

15  See for example R. Cole Harris, "The Simplification of Europe Overseas," *Annals of the Association of American Geographers* 67 (1977): 469-83, which identifies a strong sense of family-centred independence among new world settlers, and John C. Weaver, *The Great Land Rush and the Making of the Modern World, 1650-1900* (Montreal/Kingston: McGill-Queen's University Press, 2003).

16  Palmer, "Radical Reasoning."

17  Allan Smith, "The Myth of the Self-made Man in English Canada, 1850-1914," *Canadian Historical Review* 59, 2 (1978): 189-219. Elements of the argument challenged by Smith have come to underpin some of the critiques of McKay's Liberal Order Framework: see especially Jerry Bannister, "Canada as Counter-Revolution: The Loyalist Order Framework in Canadian History, 1750-1840," in *Liberalism and Hegemony*, ed. Ducharme and Constant, 64-97.

18  Smith, "Myth of the Self-made Man," 189

19  Seymour Martin Lipset, *The First New Nation: The United States in Historical and Comparative Perspective* (Garden City, NY: Doubleday, 1967), 287-88; Smith, "Myth of the Self-made Man," 190.

20  Vernon C. Fowke, "The Myth of the Self-Sufficient Canadian Pioneer," Royal Society of Canada, *Transactions* LIV, Series III (June 1962): 24; Smith, "Myth of the Self-made Man," 212-15.

21  Catherine Parr Traill, *Canadian Settler's Guide* (Toronto, Canada West: The Old Countryman Office, 1855), 1. For more on this theme, see Graeme Wynn, "Ontario: A Fine Country not Half like England," *The Canadian Geographer* 46, 4 (2002): 371-79.

22  Fecteau, "Towards a Theory of Possible History?"

23  Jean-Marie Fecteau, *La liberté du pauvre, crime et pauvreté au XIXe siècle québécois* (Montréal: vlb éditeur, 2004).

24 By scholars as diverse as F.J. Turner, *The Frontier in American History* (New York: Henry Holt, 1921), and James T. Lemon, *The Best Poor Man's Country: A Geographical Study of Early Southeastern Pennsylvania* (Baltimore: Johns Hopkins University Press, 1972).

25 Stuart Blumin, *The Urban Threshold: Growth and Change in a Nineteenth Century Community* (Chicago: University of Chicago Press, 1976), 165. However, Darren Ferry, *Uniting in Measures of Common Good: The Construction of Liberal Identities in Central Canada* (Montreal/Kingston: McGill-Queen's University Press, 2008) studies a half-dozen voluntary associations in Ontario and Quebec and accords them a critical role in the construction of a liberal social order.

26 Joseph Carens, "Possessive Individualism and Democratic Theory: Macpherson's Legacy," in J. Carens, ed., *Democracy and Possessive Individualism: The Intellectual Legacy of C.B. Macpherson* (Buffalo: SUNY Press, 1993), 3. See also David Morrice, "C.B. Macpherson's Critique of Liberal Democracy and Capitalism," *Political Studies* XLII (1994): 646-61.

27 C.B. Macpherson, "Elegant Tombstones: A Note on Friedman's Freedom," in *Democratic Theory: Essays in Retrieval,* ed. C.B. Macpherson (Oxford: Clarendon Press, 1973): Essay VII. Originally published in *Canadian Journal of Political Science/Revue canadienne de science politique* 1 (1968): 95-106.

## INTRODUCTION

1 Archives of Manitoba (hereafter AM), Manitoba, Department of Public Works, Minister's Office Files, GR 1607, G 7965, Alex Ingram to the Minister of Public Works, 17 May 1889.

2 Edward Ledohowski, *The Heritage Landscape of the Crow Wing Study Region of Southeastern Manitoba* (Winnipeg: Manitoba Culture, Heritage and Tourism, 2003), 14.

3 William Carlyle, "Agricultural Drainage in Manitoba: The Search for Administrative Boundaries," in *River Basin Management: Canadian Experiences,* ed. Bruce Mitchell and James S. Gardner (Waterloo: University of Waterloo, 1983), 279-95; William Carlyle, "Water in the Red River Valley of the North," *Geographical Review* 74, 3 (1984): 331-58; H. Albert Hochbaum, "Contemporary Drainage within the True Prairie of the Glacial Lake Agassiz Basin," in *Life, Land, and Water: Proceedings from a Conference on the Environmental Studies of the Glacial Lake Agassiz Region,* ed. William Mayer-Oakes (Winnipeg: University of Manitoba Press, 1967), 197-204; John Warkentin, "Human History of the Glacial Lake Agassiz Region in the 19th Century," in *Life, Land, and Water: Proceedings from a Conference on the Environmental Studies of the Glacial Lake Agassiz Region,* ed. William Mayer-Oakes (Winnipeg: University of Manitoba Press, 1967), 325-37; John Warkentin, "Water and Adaptive Strategies in Settling the Canadian West," *Manitoba Historical Society Transactions* 3 (1971-72): 59-73.

4 Aly M. Shady, ed., *Irrigation, Drainage, and Flood Control in Canada* (Ottawa: Irrigation Sector, Canadian International Development Agency, 1989), 9.

5 Kenneth H. Norrie, "The National Policy and the Rate of Prairie Settlement: A Review," in *The Prairie West: Historical Readings,* ed. R. Douglas Francis and Howard Palmer (Edmonton: University of Alberta Press, 1992), 248-49.

6 John Richards and Larry Pratt, *Prairie Capitalism: Power and Influence in the New West* (Toronto: McClelland and Stewart, 1979); Larry Pratt, "The State and Province-Building: Alberta's Development Strategy," in *The Canadian State: Political Economy and Political Power,* ed. Leo Panitch (Toronto: University of Toronto Press, 1977), 133-62.

7 G.R. Brooks and E. Nielsen, "Canadian Landform Examples - 40 - Red River, Red River Valley, Manitoba," *Canadian Geographer* 44 (2000): 306-11.

8 Leslie Hewes, "The Northern Wet Prairie of the United States: Nature, Sources of Information, and Extent," *Annals of the Association of American Geographers* 41, 4 (1951): 307; Hugh Prince, *Wetlands of the American Midwest: A Historical Geography of Changing Attitudes* (Chicago: University of Chicago Press, 1997), 27-74.

9 Edward Schiappa, "Towards a Pragmatic Approach to Definition: 'Wetlands' and the Politics of Meaning," in *Environmental Pragmatism,* ed. Andrew Light and Eric Katz (London: Routledge, 1996), 209-30.

10 Prince, *Wetlands of the American Midwest,* 337-47; Ann Vileisis, *Discovering the Unknown Landscape: A History of America's Wetlands* (Washington, DC: Island Press, 1997). There are ecological similarities between the wet prairie and the Prairie Lake Region, which archaeologist Scott Anfinson locates in southwestern Minnesota, eastern South Dakota, and north-central Iowa. Scott Anfinson, "Prairie, Lakes, and People: The Archaeology of Southwestern Minnesota," Rural and Regional Essay Series of the Society for the Study of Local and Regional History at the History Center, Southwest State University, Marshall, MN, 1999, 1. Recent scholarship on wetlands that addresses the American portion of the wet prairie region includes Anthony E. Carlson, "'Drain the Swamps for Health and Home': Wetlands Drainage, Land Conservation, and National Water Policy, 1850-1917" (PhD diss., University of Oklahoma, 2010), and Anthony E. Carlson, "The Other Kind of Reclamation: Wetlands Drainage and National Water Policy, 1902-1912," *Agricultural History* 84, 4 (2010): 451-78.

11 James C. Malin, *History and Ecology: Studies of the Grassland,* ed. Robert P. Swierenga (Lincoln: University of Nebraska Press, 1984).

12 W.A. Mackintosh, *Prairie Settlement: The Geographical Setting* (Toronto: Macmillan, 1934), 4-10.

13 Gene Krenz and Jay Leitch, *A River Runs North: Managing an International River* (n.p.: Red River Water Resources Council, 1993), 3.

14 Geoffrey A.J. Scott, "Manitoba's Ecoclimatic Regions," in *The Geography of Manitoba: Its Land and Its People,* ed. John Welsted, John Everitt, and Christoph Stadel (Winnipeg: University of Manitoba Press, 1997), 46; Celina Campbell et al., "Bison Extirpation May Have Caused Aspen Expansion in Western Canada," *Ecography* 17, 4 (1994): 360-62.

15 Danny Blair, "The Climate of Manitoba," in *The Geography of Manitoba: Its Land and Its People,* ed. John Welsted, John Everitt, and Christoph Stadel (Winnipeg: University of Manitoba Press, 1997), 31-39. For a discussion of the variability of Manitoba's rivers, see P. Ashmore and M. Church, *The Impact of Climate Change on Rivers and River Processes in Canada,* Geological Survey of Canada Bulletin 555 (Ottawa: Geological Survey of Canada, 2001), 31. For information on a similarly variable landscape in the United States, see Anthony J. Amato, "A Wet and Dry Landscape," in *Draining the Great Oasis: An Environmental History of Murray County, Minnesota,* ed. Anthony J. Amato, Janet Timmerman, and Joseph J. Amato (Marshall, MN: Crossings Press, 2001), 1-20.

16 Thomas E. Weber, "On Being Downstream from Everyone Else," *Canadian Water Resources Journal* 4, 3 (1979): 75-81.

17 A.G. van der Valk and C.B. Davis, "The Role of Seed Banks in the Vegetation Dynamics of Prairie Glacial Marshes," *Ecology* 59, 2 (1978): 322-35. For an alternative view on drivers of wetland ecological processes, see L.G. Goldsborough and G.G.C. Robinson, "Pattern

in Wetlands," in *Algal Ecology in Freshwater Benthic Ecosystems,* ed. R.J. Stevenson, M.L. Bothwell, and R.L. Lowe (San Diego: Academic Press, 1996), 77-117.

18 J.H. Ellis, *The Soils of Manitoba* (Winnipeg: Manitoba Economic Survey Board, 1938), 47.

19 The history of flooding in Winnipeg and southern Manitoba has been analyzed by historian J.M. Bumsted. See, for example, J.M. Bumsted, *Floods of the Centuries: A History of Flood Disasters in the Red River Valley, 1776-1997* (Winnipeg: Great Plains Publications, 1997); J.M. Bumsted, "The Manitoba Royal Commission on Flood Cost Benefit and the Origins of Cost Benefit Analysis in Canada," *American Review of Canadian Studies* 32, 1 (2002): 97-122; and J.R. Bumsted, "Flooding in the Red River Valley of the North," in *Harm's Way: Disasters in Western Canada,* ed. Anthony Rasporich and Max Foran (Calgary: University of Calgary Press, 2004), 239-63. The variation in the second initial is presumably an error.

20 For a selection of good articles as well as a general introduction to the early state formation literature, see Allan Greer and Ian Radforth, eds., *Colonial Leviathan: State Formation in Mid-Nineteenth-Century Canada* (Toronto: University of Toronto Press, 1992).

21 Bruce Curtis, *The Politics of Population: State Formation, Statistics, and the Census of Canada, 1840-1875* (Toronto: University of Toronto Press, 2001).

22 Tina Loo, *Making Law, Order, and Authority in British Columbia, 1821-1871* (Toronto: University of Toronto Press, 1994).

23 Tina Loo, *States of Nature: Conserving Canada's Wildlife in the Twentieth Century* (Vancouver: UBC Press, 2006).

24 Donald J. Pisani, *Water and American Government: The Reclamation Bureau, National Water Policy, and the West, 1902-1935* (Berkeley: University of California Press, 2002), 292.

25 Karl A. Wittfogel, *Oriental Despotism: A Comparative Study of Total Power* (New Haven: Yale University Press, 1957), 3.

26 A useful discussion of this scholarship is B.W. Kang, "The Role of Irrigation in State Formation in Ancient Korea," in *A History of Water, Volume 1: Water Control and River Biographies,* ed. Terje Tvedt and Eva Jakobsson (London: I.B. Tauris, 2006), 234-51. For a recent contribution to this literature, see Jamie Linton, *What Is Water? The History of a Modern Abstraction* (Vancouver: UBC Press, 2010), especially Chapter 3, "Intimations of Modern Water," 47-72.

27 Donald Worster, *Rivers of Empire: Water, Aridity, and the Growth of the American West* (New York: Pantheon Books, 1985), 7.

28 Fernande Roy, *Progrès, harmonie, liberté: Le libéralisme des milieux d'affaires francophones à Montréal au tournant du siècle* (Montréal: Boréal, 1988); Barry Ferguson, *Remaking Liberalism: The Intellectual Legacy of Adam Shortt, O.D. Skelton, W.C. Clark, and W.A. Mackintosh, 1890-1925* (Montreal: McGill-Queen's University Press, 1993); Loo, *Making Law.*

29 Ian McKay, "The Liberal Order Framework: A Prospectus for a Reconnaissance of Canadian History," *Canadian Historical Review* 81, 4 (2000): 617-45; Ruth Sandwell, "The Limits of Liberalism: The Liberal Reconnaissance and the History of the Family in Canada," *Canadian Historical Review* 84, 3 (2003): 423-50; Andrew Smith, "Toryism, Classical Liberalism, and Capitalism: The Politics of Taxation and the Struggle for Canadian Confederation," *Canadian Historical Review* 89, 1 (2008): 1-25. See also the recent volume reprinting McKay's essay along with other essays on his liberal order framework in Jean-François Constant and Michel Ducharme, eds., *Liberalism and Hegemony: Debating the Canadian Liberal Revolution* (Toronto: University of Toronto Press, 2009).

30  Ruth Sandwell, *Contesting Rural Space: Land Policy and the Practices of Resettlement on Saltspring Island, 1859-1891* (Montreal: McGill-Queen's University Press, 2005); James Murton, *Creating a Modern Countryside: Liberalism and Land Resettlement in British Columbia* (Vancouver: UBC Press, 2007); Catherine Anne Wilson, *Tenants in Time: Family Strategies, Land, and Liberalism in Upper Canada, 1799-1871* (Montreal: McGill-Queen's University Press, 2009); Jarett Rudy, *Freedom to Smoke: Tobacco Consumption and Identity* (Montreal: McGill-Queen's University Press, 2005); Daniel Samson, *The Spirit of Industry and Improvement: Liberal Government and Rural-Industrial Society, Nova Scotia, 1790-1862* (Montreal: McGill-Queen's University Press, 2008). Also relevant is Keith D. Smith, *Liberalism, Surveillance, and Resistance: Indigenous Communities in Western Canada, 1877-1927* (Edmonton: Athabasca University Press, 2009).

31  On the relationship between environmental history and liberalism, see also Stéphane Castonguay and Darin Kinsey, "The Nature of the Liberal Order: State Formation, Conservation, and the Government of Non-Humans in Canada," in *Liberalism and Hegemony: Debating the Canadian Liberal Revolution,* ed. Jean-François Constant and Michel Ducharme (Toronto: University of Toronto Press, 2009), 221-45.

32  In his work on farm credit in Canada, W.T. Easterbrook included drainage as among the supports that provincial governments offered to aspiring farmers. In Manitoba, Easterbrook explains, the "improvement of farm lands has involved substantial governmental assistance." W.T. Easterbrook, *Farm Credit in Canada* (Toronto: University of Toronto Press, 1938), 96.

33  There are connections here to the concept of colloquial meteorology proposed by environmental historian Liza Piper in "Colloquial Meteorology," in *Method and Meaning in Canadian Environmental History,* ed. Alan MacEachern and William J. Turkel (Toronto: Thompson-Nelson, 2008), 102-23.

34  Mark Fiege, *Irrigated Eden: The Making of an Agricultural Landscape in the American West* (Seattle: University of Washington Press, 1999); Mark Fiege, "Private Property and the Ecological Common in the American West," in *Everyday America: Cultural Landscape Studies after J.B. Jackson,* ed. Chris Wilson and Paul Groth (Berkeley: University of California Press, 2003), 219-31, 343-46; Mark Fiege, "The Weedy West: Mobile Nature, Boundaries, and Common Space in the Montana Landscape," *Western Historical Quarterly* 35, 1 (2005): 22-47. On mobile nature, see also Robert M. Wilson, *Seeking Refuge: Birds and Landscapes of the Pacific Flyway* (Seattle: University of Washington Press, 2010).

35  Dan Flores, "Place: An Argument for Bioregional History," *Environmental History Review* 18 (1994): 1-18.

36  William G. Robbins, "Bioregional and Cultural Meaning: The Problem with the Pacific Northwest," *Oregon Historical Quarterly* 103, 4 (2002): 421. The phrase "natural nations" is derived from personal correspondence between William Robbins and Michael Egan.

37  Representative examples include Warren Upham, *The Glacial Lake Agassiz,* Monographs of the United States Geological Survey 25 (Washington, DC: United States Geological Survey, 1895); John Perry Pritchet, *The Red River Valley, 1811-1849: A Regional Study* (New Haven: Yale University Press, 1942); Vera Kelsey, *Red River Runs North!* (New York: Harper, 1951); Stanley Norman Murray, "A History of Agriculture in the Valley of the Red River of the North, 1812 to 1920" (PhD diss., University of Wisconsin, 1963); and Carlyle, "Water in the Red River Valley of the North."

38  James Scott, *Seeing like a State: Why Certain Schemes to Improve the Human Condition Have Failed* (New Haven: Yale University Press, 1998).

CHAPTER 1: DRAINS AND CULTURAL COMMUNITIES

1 Douglas Owram, *Promise of Eden: The Canadian Expansionist Movement and the Idea of the West, 1856-1900* (Toronto: University of Toronto Press, 1980), 61-62; Suzanne Zeller, *Inventing Canada: Early Victorian Science and the Idea of a Transcontinental Nation* (Toronto: University of Toronto Press, 1987), 97.

2 Henry Youle Hind, *Reports on the North-West Territory* (Toronto: John Lovell, 1859), 14.

3 R.G. Macbeth, *The Selkirk Settlers in Real Life* (Toronto: W. Briggs, 1897), 45-47; Barry Kaye, "'The Settlers' Grand Difficulty': Haying in the Economy of the Red River Settlement," *Prairie Forum* 9, 1 (1984): 1; Shannon Stunden Bower, "The Great Transformation? Wetlands and Land Use in Late 19th Century Manitoba," *Journal of the Canadian Historical Association* 15, 1 (2004): 29-47.

4 Useful comparisons of the Canadian and American experience include Donald Worster, "Two Faces West: The Development Myth in Canada and the United States," in *Terra Pacifica: People and Place in the Northwest States and Western Canada*, ed. Paul W. Hirt (Pullman: Washington State University Press, 1998), 71-91; and Paul W. Gates (with Lillian F. Gates), "Canadian and American Land Policy Decisions, 1930," in *The Jeffersonian Dream: Studies in the History of American Land Policy and Development*, ed. Allan G. Bogue and Margaret Beattie Bogue (Albuquerque: University of New Mexico Press, 1996), 148-65.

5 John C. Weaver, *The Great Land Rush and the Making of the Modern World, 1650-1900* (Montreal: McGill-Queen's University Press, 2003), 350.

6 Kirk N. Lambrecht, *The Administration of Dominion Lands, 1870-1930* (Winnipeg: Hignell Printing, 1991), 20; John Langton Tyman, *By Section, Township, and Range: Studies in Prairie Settlement* (Brandon: Assiniboine Historical Society, 1972), 12-13. The classic study is Chester Martin, *"Dominion Lands" Policy* (Toronto: Macmillan, 1938); note especially Chapter 9, "The Free Homestead for 'Dominion Lands.'"

7 A useful source on public works across Canada is Jonathan F. Vance, *Building Canada: People and Projects that Shaped the Nation* (Toronto: Penguin, 2006). Especially relevant here is his chapter on roads titled "A Unity Based on Road and Wheel," 28-51.

8 Gerald Friesen (co-written with Jean Friesen), "River Road," in *River Road: Essays on Manitoba and Prairie History* (Winnipeg: University of Manitoba Press, 1996), 3-12.

9 AM, William Pearce Fonds, MG 9 A40, 156.

10 AM, Manitoba, Department of Public Works, Minister's Office Files, GR 1607, G 7972, item 115, J.A. Macdonell to the Minister of Public Works, 25 January 1894.

11 AM, Manitoba, Department of Public Works, Deputy Minister's Office Files, GR 1610, reel M 951, H.A. Bowman, Assistant Deputy Minister, to W.H. Montague, Minister of Public Works, 13 December 1913.

12 AM, Lowe Farm Fonds, MG 8 A16, Lucinda Westover to Flora Davidson, 15 August 1880.

13 Lambrecht, *The Administration of Dominion Lands,* 12.

14 AM, Manitoba, Drainage Maintenance Boards, Minutes and Office Files, GR 1617, G 5324, F.E. Umphrey to Harry Christopherson, 14 March 1947.

15 AM, Manitoba, Legislative Assembly, Unpublished Sessional Papers, GR 174, G 8303, Return to an Order (No. 56) re Correspondence re Transfer of Natural Resources from the Dominion to Provincial Government. See also AM, Manitoba, Premier's Office Files (Norquay Administration), GR 553, Letterbook D, 35-39, J. Norquay to Charles Brodie, 14 July 1883. In this letter, Norquay explains that Manitoba has "taken no active part in

trying to induce immigration, for the very reason that immigration increases our expenses, without contributing anything to our revenue."

16  AM, Manitoba, Department of Public Works, Minister's Office Files, GR 1609, G 8046, file Special Drainage Survey Conducted by M.A. Lyons, Lyons to Errick F. Willis, Minister of Public Works, 6 May 1948.

17  John Friesen, "Expansion of Settlement in Manitoba," in *Historical Essays on the Prairie Provinces,* ed. Donald Swainson (Toronto: McClelland and Stewart, 1970), 120-30; John Warkentin, "Water and Adaptive Strategies in Settling the Canadian West," *Manitoba Historical Society Transactions* 3 (1971-72): 59-73.

18  "Draining Lands," *The Emigrant,* 1 July 1886, 34. For a later illustration of this, see "Towns, Cities, Villages, and Rural Municipalities of the Great West," *Western Municipal News,* October 1906, 219.

19  John C. Lehr, "Settlement: The Making of a Landscape," in *The Geography of Manitoba: Its Land and Its People,* ed. John Welsted, John Everitt, and Christoph Stadel (Winnipeg: University of Manitoba Press, 1997), 97.

20  J.H. Ellis, *The Soils of Manitoba* (Winnipeg: Manitoba Economic Survey Board, 1938), 27-29.

21  AM, Manitoba, Department of Public Works, Deputy Minister's Office Files, GR 1611, G 8062, file Reclamation Branch, Chief Engineer Bowman to D.G. McKenzie, Acting Minister of Public Works, 12 April 1929. See also AM, GR 1609, G 8046, file Union of Municipal Drainage Maintenance Districts, text of an address delivered at the annual meeting of the Union of Municipal Drainage Districts, 27 November 1945. In this speech, F.E. Umphrey explained how "during the dry years land that was broken up for cultivation ... was not, however, returned to meadow or pasture when the wet seasons returned, but instead, the drainage system was extended, thus laying the foundation for the complex problem we have on our hands today."

22  AM, GR 1610, reel M 951, H.A. Bowman, Assistant Deputy Minister, to W.H. Montague, Minister of Public Works, 13 December 1913.

23  AM, GR 1607, G 7979, item 566, G.S. Howard to Robert Watson, Minister of Public Works, Petition re Drainage in Cromwell District, 15 May 1896.

24  Luna B. Leopold, *Water: A Primer* (San Francisco: W.H. Freeman, 1974), 63.

25  Warkentin, "Water and Adaptive Strategies in Settling the Canadian West," 60. As Warkentin explains, "in the 1870s a few surface drains were dug, usually in connection with road or railroad building. In 1880, the Manitoba Department of Public Works started the attempt to drain the land between the Escarpment and the Red, by constructing a few ditches through the marshes. Accomplishments were limited, because comprehensive draining schemes extending over large areas and including many miles of lateral ditches were what was needed, not piecemeal efforts."

26  AM, Manitoba, Executive Council Office, Clerk's Office, Orders in Council, GR 1530, Order in Council 619, Report of the Minister of Public Works on Lands [illegible] Drainage, 14 January 1882.

27  AM, GR 1530, Order in Council 213, Re Drainage of Wet Lands Report Sent to Sec. State, 28 May 1880.

28  AM, Manitoba, Executive Council Office, Clerk's Office, Minutes, GR 1659, Order in Council 6, Minutes of the Executive Council, 21 February 1877.

29  Owram, *Promise of Eden,* 171.

30 Gerald Friesen, *The Canadian Prairies: A History* (Toronto: University of Toronto Press, 1984), 218.

31 AM, Manitoba, Executive Council Office, Premier's Office Files (Greenway Administration), GR 1662, G 500, item 5009, Robert Wemyss to Thomas Greenway, Premier of Manitoba, 10 August 1892.

32 AM, GR 1607, G 7986, item 1434, Oswald Berire to R. Watson, 3 September 1898. For another example, see AM, GR 1609, G 8011, item 521, Alf. Douglas to Mr. Hastings, 23 January 1903.

33 AM, GR 1607, G 7970, item 17 1/2, Petition re Municipality of Lansdowne, Grassy River Drain, n.d.

34 AM, GR 1607, G 7962, Thos. Usher to Wilson, Department of Public Works, 13 July 1887.

35 AM, GR 1607, G 7961, J.O. Smith to Wade, Department of Public Works, 10 November 1886.

36 This definition is adapted from Engin F. Isin, "The Origins of Canadian Municipal Government," in *Canadian Metropolitics: Governing Our Cities,* ed. James Lightbody (Toronto: Copp Clark, 1995), 51. For a good recent discussion of municipal government in the context of the liberal order framework, see Michèle Dagenais, "The Municipal Territory: A Product of the Liberal Order?" in *Liberalism and Hegemony: Debating the Canadian Liberal Revolution,* ed. Jean-François Constant and Michel Ducharme (Montreal: McGill-Queen's University Press, 2009), 201-20.

37 AM, GR 1607, G 7985, item 659, C.A. Millican to Minister of Public Works, 14 April 1898.

38 Lenore Eidse, ed., *Furrows in the Valley: Rural Municipality of Morris, 1880-1980* (n.p.: Inter-Collegiate Press, 1980), 30-31.

39 For examples of public controversy elsewhere, see Mark Fiege, *Irrigated Eden: The Making of an Agricultural Landscape in the American West* (Seattle: University of Washington Press, 1999).

40 Mary R. McCorvie and Christopher L. Lant, "Drainage District Formation and the Loss of Midwestern Wetlands, 1850-1930," *Agricultural History* 67, 4 (1993): 26. Drainage districts can usefully be considered in relation to a body of literature dealing with special government districts. Although most scholars have focused on later districts conceived for reasons other than drainage, this literature helps to situate the drainage district within an institutional lineage. Unfortunately, to date there has been little attention to special government districts in Canada. Works on special government districts elsewhere include John C. Bollens, *Special District Governments in the United States* (Berkeley: University of California Press, 1957); Nancy Burns, *The Formation of American Local Governments: Private Values in Public Institutions* (New York: Oxford University Press, 1994); Kathryn A. Foster, *The Political Economy of Special-Purpose Government* (Washington: Georgetown University Press, 1997); and Donald Foster Stetzer, "Special Districts in Cook County: Toward a Geography of Local Government," Department of Geography, University of Chicago, Research Paper 169, 1975.

41 J.A. Griffiths, "The History and Organization of Surface Drainage in Manitoba," paper presented before the Winnipeg Branch, Engineering Institute of Canada, 20 March 1952, Sciences and Engineering Library, University of Manitoba, 7.

42 Friesen, *The Canadian Prairies,* 244-45.

43 Gerald Friesen, "Perimeter Vision: Three Notes on the History of Rural Manitoba," in *River Road: Essays on Manitoba and Prairie History* (Winnipeg: University of Manitoba

Press, 1996), 197-214; Thomas Peterson, "Manitoba: Ethnic and Class Politics," in *Canadian Provincial Politics: The Party Systems of the Ten Provinces,* 2nd ed., ed. Martin Robin (Scarborough: Prentice-Hall of Canada, 1978), 61-119; Nelson Wiseman, "The Pattern of Prairie Politics," in *The Prairie West: Historical Readings,* 2nd ed., ed. R. Douglas Francis and Howard Palmer (Edmonton: University of Alberta Press, 1992), 640-60.

44 Robert Murchie, *Agricultural Progress on the Prairie Frontier* (Toronto: Macmillan, 1936), 141-42.

45 On ethnicity in Manitoba history, see Friesen, *The Canadian Prairies;* Yossi Katz and John Lehr, *The Last Best West: Essays on the Historical Geography of the Canadian Prairies* (Jerusalem: Magnes Press, 1999); Frances Swyripa, *Storied Landscapes: Ethno-Religious Identity and the Canadian Prairies* (Winnipeg: University of Manitoba Press, 2010); and Royden Loewen, *Family, Church, and Market: A Mennonite Community in the Old and the New Worlds, 1850-1930* (Urbana: University of Illinois Press, 1993). A focus on the forces that cut across ethnicity is provided in John C. Lehr and Yossi Katz, "Crown, Corporation, and Church: The Role of Institutions in the Stability of Pioneer Settlements in the Canadian West, 1870-1914," *Journal of Historical Geography* 21, 4 (1995): 413-29. A recent addition to this scholarship with a focus on the urban context is Royden Loewen and Gerald Friesen, *Immigrants in Prairie Cities: Ethnic Diversity in Twentieth-Century Canada* (Toronto: University of Toronto Press, 2009).

46 J.M.S. Careless, "'Limited Identities' in Canada," *Canadian Historical Review* 50, 1 (1969): 1-10.

47 Gerald Friesen (co-written with Royden Loewen), "Romantics, Pluralists, Postmodernists: Writing Ethnic History in Prairie Canada," in *River Road: Essays on Manitoba and Prairie Culture* (Winnipeg: University of Manitoba Press, 1996), 183-96.

48 Adolf Ens, *Subjects or Citizens? The Mennonite Experience in Canada, 1870-1925* (Ottawa: University of Ottawa Press, 1994), 3.

49 Ibid., 7.

50 John Warkentin, "The Mennonite Settlements of Southern Manitoba" (PhD diss., University of Toronto, 1960), 38.

51 Ibid., 1.

52 Ibid., 59.

53 Loewen, *Family, Church, and Market,* 80-81.

54 Warkentin, "The Mennonite Settlements of Southern Manitoba," 38.

55 Loewen, *Family, Church, and Market,* 107-15.

56 Ens, *Subjects or Citizens?,* 35.

57 Warkentin, "The Mennonite Settlements of Southern Manitoba," 43.

58 Ibid., 302.

59 Regina (Doerksen) Neufeld, "Schantzenberg," in *Working Papers of the East Reserve Village Histories, 1874-1910,* ed. John Dyck (Steinbach, MB: Hanover Steinbach Historical Society, 1990), 101.

60 Gerhard Ens, *Volost and Municipality: The Rural Municipality of Rhineland, 1884-1984* (Altona, MB: Friesen Printers, 1984), 22; Warkentin, "The Mennonite Settlements of Southern Manitoba," 46-47.

61 Ens, *Volost and Municipality,* 74.

62 Ibid.

63 AM, GR 1607, G 7974, item 932, C. Hiebert, Municipality of Rhineland Secretary-Treasurer, to the Minister of Public Works, 11 July 1894.

64 Ens, *Volost and Municipality,* 74.

65  Statistics Canada, Table XI, "Origins of the People," 1901 census.

66  *Morning Telegram*, 12 March 1903.

67  Royden Loewen, *Diaspora in the Countryside: Two Mennonite Communities and Mid-Twentieth-Century Rural Disjuncture* (Toronto: University of Toronto Press, 2006), 69.

68  AM, GR 1530, OC 8466.

69  Eidse, ed., *Furrows in the Valley*, 31. The organization is identified in this community history as the Red River Valley Drainage Improvement Association (note the omission of "and," which appears to have been an error). See also AM, GR 1609, G 8019, file Drainage – General, 1919-1927, Drainage Commission, Notes by Mr. McColl.

70  Adapted from Statistics Canada, Table 27, "Population Classified According to Principal Origin of the People by Counties and Their Subdivision," 1921 census. The percentages do not add up to 100 since ethnicities not within the top three for any of the relevant rural municipalities were excluded. Mennonites were identified as Dutch, rather than as German, due to lingering wartime prejudices.

71  W.J. Carlyle, "Mennonite Agriculture in Manitoba," *Canadian Ethnic Studies* 13, 2 (1981): 72-97.

72  Royden Loewen, "Ethnic Farmers and the 'Outside' World: Mennonites in Manitoba and Nebraska, 1874-1900," *Journal of the Canadian Historical Association* 1 (1990): 195-213, especially 204-5. See also Loewen, *Family, Church, and Market,* Chapter 7, "Market Farming and the Mennonite Household," 125-49.

73  Arthur Ray, *Indians in the Fur Trade: Their Role as Trappers, Hunters, and Middlemen in the Lands Southwest of Hudson Bay, 1660-1870* (Toronto: University of Toronto Press, 1998), 3-5, 187.

74  Laura Peers, *The Ojibwa of Western Canada, 1780 to 1870* (Winnipeg: University of Manitoba Press, 1994), 3-26.

75  George Van Der Goes Ladd, *Shall We Gather at the River?* (Toronto: United Church of Canada, 1986), 23.

76  Peers, *The Ojibwa of Western Canada,* 198.

77  Jean Friesen, "Grant Me Wherewith to Make My Living," in *Aboriginal Resource Use in Canada: Historical and Legal Aspects,* ed. Kerry Abel and Jean Friesen (Winnipeg: University of Manitoba Press, 1991), 141-43.

78  Ibid., 141.

79  Arthur J. Ray, Jim Miller, and Frank J. Tough, *Bounty and Benevolence: A History of Saskatchewan Treaties* (Montreal: McGill-Queen's University Press, 2000), 65.

80  Alexander Morris, *The Treaties of Canada* (Toronto: Belfords, Clarke and Company, 1880), 28-34.

81  Friesen, "Grant Me Wherewith to Make My Living," 152.

82  Library and Archives Canada (hereafter LAC), Canada, Department of Indian Affairs Records, RG 10, vol. 6615, file 7125-5, reel C-8018, H.D. Latulippe, Indian Agent, to Secretary, Department of Indian Affairs, 22 June 1920; LAC, RG 10, vol. 6615, file 7125-4, reel C-8018, Memorandum for Deputy Minister by D. Laird, Indian Commissioner, 7 February 1911.

83  AM, Manitoba, Department of Natural Resources, Miscellaneous Land Files, GR 7700, reel M 1639, Drainage Districts General File, James A. Gardiner to R.S. Moore, 26 February 1949.

84  LAC, RG 10, vol. 7592, file 7125-6 part 1, reel C-11564, McColl to Inspector General of Indian Affairs, 4 July 1886.

85  Ibid.

86  Sarah Carter, *Lost Harvests: Prairie Indian Reserve Farmers and Government Policy* (Montreal: McGill-Queen's University Press, 1990), 244; Sarah Carter, "'An Infamous Proposal': Prairie Indian Reserve Land and Soldier Settlement after World War I," *Manitoba History* 37 (1999): 9-21.

87  T.C.B. Boon, "St. Peter's Dynevor, the Original Indian Settlement of Western Canada," *Manitoba Historical Society Transactions* 3 (1952-53): 16-32; Canada, "Peguis First Nation Inquiry Treaty Land Entitlement Claim," in *Indian Claims Commission Proceedings,* vol. 14 (Ottawa: Indian Claims Commission, 2001), 22-30.

88  LAC, RG 10, vol. 10303, file 501/8-4-11, reel T-7579, Report from Indian Agent, Fisher River Agency, to Assistant Deputy and Secretary, Department of Indian Affairs, 22 January 1914.

89  Carter, *Lost Harvests,* 209-36.

90  See also Leo G. Waisberg and Tim E. Holzkamm, "A Tendency to Discourage Them from Cultivating: Ojibwa Agriculture and Indian Affairs Administration in Northwestern Ontario," *Ethnohistory* 40, 2 (1993): 175-211.

91  LAC, RG 10, vol. 8313, file 577/8-4-3-6 1, reel C-13783, Memorandum, 27 June 1906.

92  LAC, RG 10, vol. 8313, file 577/8-4-3-6 1, reel C-13783, Inspector of Indian Agencies Marlatt to Commissioner of Indian Affairs Laird, 29 June 1906.

93  For a related argument, see Donald J. Pisani, "The Dilemmas of Indian Water Policy, 1887-1928," in *Fluid Arguments: Five Centuries of Western Water Conflict,* ed. Char Miller (Tucson: University of Arizona Press, 2001), 78-94.

## CHAPTER 2: JURISDICTIONAL QUAGMIRES

1  W.J. Mitsch and G.W. Gosselink, *Wetlands,* 3rd ed. (New York: John Wiley and Sons, 2000), 3-23, 649.

2  W.L. Morton, "Introduction," in *Manitoba: The Birth of a Province* (Winnipeg: Manitoba Record Society Publications, 1984), xxiv-xxv.

3  Jim Mochoruk, *Formidable Heritage: Manitoba's North and the Cost of Development 1870 to 1930* (Winnipeg: University of Manitoba Press, 2004), 232-33.

4  Gerald Friesen, *The Canadian Prairies: A History* (Toronto: University of Toronto Press, 1984), 162-94.

5  Mochoruk, *Formidable Heritage,* 105-51.

6  G. Mulamoottil, B.G. Warner, and E.A. McBean, "Introduction," in *Wetlands: Environmental Gradients, Boundaries, and Buffers,* ed. George Mulamoottil, Barry G. Warner, and Edward A. McBean (Boca Raton: Lewis Publishers, 1996), 5; V. Carter, "Environmental Gradients, Boundaries, and Buffers: An Overview," in *Wetlands: Environmental Gradients, Boundaries, and Buffers,* ed. George Mulamoottil, Barry G. Warner, and Edward A. McBean (Boca Raton: Lewis Publishers, 1996), 11; Edward Schiappa, "Towards a Pragmatic Approach to Definition: 'Wetlands' and the Politics of Meaning," in *Environmental Pragmatism,* ed. Andrew Light and Eric Katz (London: Routledge, 1996), 209-30.

7  Carter, "Environmental Gradients, Boundaries, and Buffers," 9.

8  Cole Harris, "How Did Colonialism Dispossess? Comments from an Edge of Empire," *Annals of the Association of American Geographers* 94, 1 (2004): 165-82; James Scott, *Seeing like a State: Why Certain Schemes to Improve the Human Condition Have Failed* (New Haven: Yale University Press, 1998), 11-52; David N. Livingstone, *Putting Science in Its Place: Geographies of Scientific Knowledge* (Chicago: University of Chicago Press, 2003); see especially the section titled "Mapping Territory," 153-63. For an article that deals with

technologies such as mapping but without particular attention to the state, see Bruno Latour, "Circulating Reference: Sampling the Soil in the Amazon Forest," in *Pandora's Hope: Essays on the Reality of Science Studies* (Cambridge, MA: Harvard University Press, 1999), 24-79. This article also addresses the difficulties of drawing boundaries on the basis of environmental characteristics such as soil type or vegetation pattern. For a recent analysis of the materialization and abstraction of water, see Jamie Linton, *What Is Water? The History of a Modern Abstraction* (Vancouver: UBC Press, 2010), especially Chapter 7, "Reading the Resource: Modern Water, the Hydrologic Cycle, and the State," 148-61.

9  Mochoruk, *Formidable Heritage,* 111.

10  Ann Vileisis, *Discovering the Unknown Landscape: A History of America's Wetlands* (Washington, DC: Island Press, 1997), 72.

11  Michael Williams, "Agricultural Impacts in Temperate Wetlands," in *Wetlands: A Threatened Landscape,* ed. Michael Williams (Oxford: Basil Blackwell, 1990), 205.

12  Margaret Beattie Bogue, "The Swamp Land Act and Wet Land Utilization in Illinois, 1850-1890," *Agricultural History* 25, 4 (1951): 169-80; Anthony E. Carlson, "'Drain the Swamps for Health and Home': Wetlands Drainage, Land Conservation, and National Water Policy, 1850-1917" (PhD diss., University of Oklahoma, 2010), 109-47; Hugh Prince, *Wetlands of the American Midwest: A Historical Geography of Changing Attitudes* (Chicago: University of Chicago Press, 1997), 140-48; Williams, "Agricultural Impacts in Temperate Wetlands," 204-5.

13  Not only is there evidence that Canadian politicians were aware of the American precedent, but also it is clear that they used the American legislation as a model. LAC, Canada, Department of the Interior, RG 15, vol. 1207, file 142200 1, H.H. Smith to A.M. Burgess, 19 August 1889.

14  AM, Manitoba, Natural Resources Department, Miscellaneous Land Files, GR 7700, reel M 1640, Swamp Lands, vol. 1, Copy of a Report of a Committee of the Privy Council, 21 April 1884.

15  AM, Manitoba, Department of Natural Resources, Miscellaneous Land Files, GR 7721, reel M 1690, file 4834, Lindsay Russell to J.S. Dennis, 24 June 1880.

16  AM, Manitoba, Executive Council Office, Clerk's Office, Orders in Council, GR 1530, Order in Council 619, Report of the Minister of Public Works on Lands Reclaimed by Drainage, 14 January 1882. AM, GR 7721, reel M 1690, file 5, Memorandum from the Minister of the Interior, 25 June 1880.

17  AM, GR 7721, reel M 1690, file 4834, J.W. Harris to Aquila Walsh, 15 January 1883.

18  AM, GR 7721, reel M 1690, file 4834, Unsigned Draft of a Letter to C.P. Brown, 15 November 1882.

19  AM, GR 7721, reel M 1690, file 4834, E. Deville, Chief Inspector of Surveys, to A.M. Burgess, Deputy Minister of the Department of the Interior, 25 August 1884.

20  AM, Manitoba, Legislative Assembly, Unpublished Sessional Papers, GR 174, G 8128, Return to an Order re Transactions of Reclaimed Lands, 1886. See also AM, GR 7721, reel M 1690, file 4834, Memo (not signed or dated).

21  AM, GR 7721, reel M 1690, file 4834, Copy of a Report of a Committee of the Privy Council, 21 April 1884.

22  AM, GR 174, G 8128, Return to an Order re Transactions of Reclaimed Lands, 1886.

23  John Langton Tyman, *By Section, Township, and Range: Studies in Prairie Settlement* (Brandon: Assiniboine Historical Society, 1972), 29. Langton asserts that 250,000 acres were reclaimed, resulting in a transfer of 112,146 acres. Although these amounts do not

correspond precisely to those cited elsewhere (see, e.g., AM, GR 174, G 8128, Return to an Order re Transactions of Reclaimed Lands, 1886), all amounts are fairly similar.

24 AM, GR 7721, Swamp Lands File, vol. 16, reel M 1690, Copy of a Report of a Committee of the Executive Council, 20 September 1906. In the same file, see also the Memo to the Governor General of Canada in Council from Premier R.P. Roblin, 5 November 1906. In the latter document, the amount of swampland is estimated at between 8 and 10 million acres.

25 AM, Manitoba, Premier's Office Files (Norquay Administration), GR 553, Letterbook D, Norquay to unnamed correspondent, 7 February 1885.

26 AM, GR 7721, reel M 1690, Swamp Lands File, vol. 16, Order of the Privy Council, PC 1037, 19 June 1886.

27 AM, GR 7700, reel M 1640, Swamp Lands File, vol. 1, Swamp Lands Commissioners Wagner and Crawford to the Provincial Lands Commissioner, 11 October 1895.

28 AM, GR 7721, reel M 1690, Swamp Lands File, vol. 16, Burgess to Larivière and Harrison, 9 December 1887, and Copy of a Report of a Committee of the Executive Council, 20 September 1906; AM, GR 7700, reel M 1640, Swamp Lands File, vol. 1, William Crawford, Swamp Lands Commissioner, to James A. Smart, Manitoba Minister of Public Works and Commissioner of Lands, 21 March 1892.

29 AM, GR 7700, reel M 1640, Swamp Lands File, vol. 1, M.A. Ferries, Provincial Lands Inspector to the Provincial Lands Commissioner, 22 November 1895; AM, GR 174, G 8241, Report of the Provincial Lands Department for 1905; LAC, Canada, Department of Justice, RG 13, vol. 2321, file 928/1904, Secretary of the Department of the Interior to the Deputy Minister of Justice, 16 November 1904.

30 LAC, Canada, Department of the Interior, RG 15, vol. 1207, file 142200 1, H.H. Smith, Commissioner of Dominion Lands, to A.M. Burgess, Deputy Minister of the Interior, 6 August 1889.

31 Mulamoottil, Warner, and McBean, "Introduction," 2.

32 AM, GR 7721, reel M 1690, Swamp Lands File, vol. 16, Memorial to the Governor General of Canada in Council from Premier R.P. Roblin, 5 November 1906.

33 AM, GR 7700, reel M 1640, Swamp Lands File, vol. 1, Provincial Lands Commissioner to Minister of the Interior, 8 April 1908.

34 AM, GR 7700, reel M 1640, Swamp Lands File, vol. 1, J. Obed Smith, Chief Clerk, Manitoba Commissioner of Railways, to the Provincial Lands Commissioner, 17 August 1896.

35 AM, GR 174, G 8165, Return to an Order of the House Showing a Copy of All Correspondence with the Dominion Government Having Reference to Swamp Lands, 1891.

36 AM, GR 174, G 8241, Report of the Provincial Lands Commissioner for 1906.

37 AM, GR 7700, reel M 1640, Swamp Lands File, vol. 1, A.M. Burgess, Deputy Minister of the Interior, to A.C. Larivière and D.H. Harrison, 9 December 1887.

38 AM, GR 7721, reel M 1690, Swamp Lands File, vol. 16, Copy of a Report of a Committee of the Executive Council, 20 September 1906.

39 Swamplands were administered under the Provincial Lands Act, which came into force on 1 July 1887. See John Langton Tyman, "Patterns of Western Land Settlement," *Manitoba Historical Society Transactions* 3 (1971-72): 117-35.

40 AM, Department of Public Works, Minister's Office Files, GR 1607, G 7977, William Wagner and William Crawford, Swamp Lands Commissioners, to Robert Watson, Minister of Public Works, 3 September 1895.

41　LAC, RG 13, vol. 80, file 143, A.M. Burgess, Deputy Minister of the Interior, to Robert Sedgewick, Deputy Minister of Justice, 4 February 1891.

42　AM, GR 7700, reel M 1640, Swamp Lands File, vol. 1, Secretary of the Department of the Interior to Clifford Sifton, Land Commissioner, 11 January 1893.

43　AM, GR 7700, reel M 1640, Swamp Lands File, vol. 1, Secretary of the Department of the Interior to Clifford Sifton, Land Commissioner, 11 January 1893. Italics in original.

44　AM, GR 7721, reel M 1690, Swamp Lands File, vol. 16, Copy of a Report of a Committee of the Executive Council, 20 September 1906. For a discussion of the controversy surrounding taxes and duties in British Columbia that suggests what was at stake in an era before income tax, see Daniel P. Marshall, "An Early Rural Revolt: The Introduction of the Canadian System of Tariffs to British Columbia, 1871-74," in *Beyond the City Limits: Rural History in British Columbia,* ed. Ruth Sandwell (Vancouver: UBC Press, 1999), 47-61.

45　Mochoruk, *Formidable Heritage,* 140-41.

46　AM, William Pearce Fonds, MG 9 A40, 40-45.

47　Tyman, *By Section, Township, and Range,* 29.

48　Roderick Irwinn Stutt, "Water Policy-Making in the Canadian Plains: Historical Factors that Influenced the Work of the Prairie Provinces Water Board (1948-1969)" (PhD diss., University of Regina, 1995), 32.

49　E.A. Alyn Mitchner, "The Development of Western Waters, 1885-1930" (unpublished manuscript, University of Alberta, 1973), 25. See also E.A. Mitchner, "William Pearce and Federal Government Activity in Western Canada, 1882-1904" (PhD diss., University of Alberta, 1971).

50　Stutt, "Water Policy-Making in the Canadian Plains," 32.

51　Mitchner, "The Development of Western Waters," 50.

52　James L. Wescoat Jr., "'Watersheds' in Regional Planning," in *The American Planning Tradition: Culture and Policy,* ed. Robert Fishman (Washington, DC: Woodrow Wilson Center Press, 2000), 147-71; Donald Worster, *A River Running West: The Life of John Wesley Powell* (New York: Oxford University Press, 2001); Wallace Earle Stegner, *Beyond the Hundredth Meridian: John Wesley Powell and the Second Opening of the West* (Lincoln: University of Nebraska Press, 1982).

53　Mitchner, "The Development of Western Waters," 52.

54　Christopher Armstrong, Matthew Evenden, and H.V. Nelles, *The River Returns: An Environmental History of the Bow* (Montreal: McGill-Queen's University Press, 2009), 153-57; C.S. Burchill, "The Origins of Canadian Irrigation Law," *Canadian Historical Review* 29, 4 (1948): 358.

55　Mitchner, "The Development of Western Waters," 72.

56　Burchill, "The Origins of Canadian Irrigation Law," 362.

57　University of Alberta Archives (hereafter UAA), William Pearce Fonds, 74-169-211, Elwood Mead to William Pearce, 13 March 1896.

58　Matthew Evenden, "Precarious Foundations: Irrigation, Environment, and Social Change in the Canadian Pacific Railway's Eastern Section, 1900-1930," *Journal of Historical Geography* 32, 1 (2006): 74-95.

59　UAA, William Pearce Fonds, 74-169-211, William Pearce to Elwood Mead, 18 March 1896. See also Armstrong, Evenden, and Nelles, *The River Returns,* especially Chapter 6, "Watering a Dry Country." For another useful discussion of the Canadian situation in relation to

American and Australian patterns, see David R. Percy, "Water Law of the Canadian West: Influences from the Western States," in *Law for the Elephant, Law for the Beaver: Essays in the Legal History of the North American West,* ed. John MacLaren, Hamar Foster, and Chet Orloff (Regina: Canadian Plains Research Center, 1992), 274-91.

60  For a suggestive discussion of the problem of seepage, see Mark Fiege, *Irrigated Eden: The Making of an Agricultural Landscape in the American West* (Seattle: University of Washington Press, 1999), 25-41.

61  LAC, Canada, Water Resources Branch, RG 89, vol. 49, file 22, General Information – Drainage, Memorandum to Mr. Rothwell, 13 January 1913.

62  LAC, Canada, Department of Public Works, RG 11, vol. 4354, file 5816-1-A to file 5816-1-C, and vol. 4355, file 5816-1-D to file 5816-1-F.

63  John Warkentin, "Water and Adaptive Strategies in Settling the Canadian West," *Manitoba Historical Society Transactions* 3 (1971-72): 67.

64  Lowe Farm Chamber of Commerce, *Lowe Farm: 75th Anniversary, 1899-1974* (Altona, MB: D.W. Friesen and Sons, 1974), 15.

65  Shannon Stunden Bower, "Natural and Unnatural Complexities: Flood Control along Manitoba's Assiniboine River," *Journal of Historical Geography* 36, 1 (2010): 57-67.

66  LAC, RG 89, vol. 290, file 4222, Reclamation Service Report, 8 December 1920.

67  LAC, RG 89, vol. 49, file 22, General Information – Drainage, Unsigned Memorandum to Mr. Roche, Minister of the Interior, 13 October 1913.

68  LAC, RG 89, vol. 49, file 22, General Information – Drainage, extract of the *Journal of the Engineering Institute of Canada,* March 1919.

69  LAC, RG 89, vol. 49, file 22, General Information – Drainage, Memorandum for F.E. Drake, 11 July 1919.

70  Ibid.

71  Ibid.

72  LAC, RG 89, vol. 50, file 22, General Information – Drainage, J.B. Harkin to F.E. Drake, 7 June 1917.

73  LAC, RG 15, D-II-I, vol. 439, Swamp Lands File, vol. 15, Memorandum to Mr. Cory, 25 July 1913.

74  LAC, RG 15, D-II-I, vol. 439, Swamp Lands File, vol. 15, Memorandum re Manitoba Swamp Lands, 6 March 1914.

75  LAC, RG 89, vol. 49, file 22, General Information – Drainage, Memorandum to Cory, 12 August 1919.

76  AM, GR 1530, Order in Council 30724, Report of a Committee of the Executive Council, 17 January 1919.

77  Scott, *Seeing like a State,* 46.

78  Edward Schiappa, "Towards a Pragmatic Approach to Definition: 'Wetlands' and the Politics of Meaning," in *Environmental Pragmatism,* ed. Andrew Light and Eric Katz (London: Routledge, 1996), 209-30. For more on the matter, see James G. Gosselink and Edward Maltby, "Wetlands Losses and Gains," in *Wetlands: A Threatened Landscape,* ed. Michael Williams (Oxford: Basil Blackwell, 1990), 296-322.

79  For a discussion of how relatively recent wetlands management differs in Canada and the United States, see John C. McLaughlin, "Progress, Politics, and the Role of Conservation: Wetland Drainage in Ontario" (PhD diss., Queen's University, 1995), 290-91. The key point is that wetlands management is primarily a provincial concern in Canada and a federal concern in the United States.

80  LAC, RG 89, vol. 290, file 4222, Memorandum for W.W. Cory by F.E. Drake, 5 November 1918.

81  Gerald Friesen, "The Prairies as Region: The Contemporary Meaning of an Old Idea," in *River Road: Essays on Manitoba and Prairie History* (Winnipeg: University of Manitoba Press, 1996), 173-74.

## CHAPTER 3: DRAINS AND GEOGRAPHICAL COMMUNITIES

1   AM, Manitoba, Department of Public Works, Minister's Office Files, GR 1609, G 8019, file Drainage – General, Drainage Committee Report, Testimony of L.C. Wilkin and Kirk, 109-16.

2   W.L. Morton, *Manitoba: A History,* 2nd ed. (Toronto: University of Toronto Press, 1967), 284.

3   John Warkentin, "Water and Adaptive Strategies in Settling the Canadian West," *Manitoba Historical Society Transactions* 3 (1971-72): 60.

4   Drainage District No. 21 was proclaimed, but significant construction was not undertaken. Although they were numbered to twenty-four, there were twenty-three functional drainage districts. Manitoba, Land Drainage Arrangement Commission, *Report of the Land Drainage Arrangement Commission Respecting Municipalities Containing Land Subject to Review under "The Land Drainage Act"* (Winnipeg: King's Printer, 1936), 55-56.

5   Saskatchewan Archives Board, Regina Branch, Coll. 276, Acc. No. 614, George Spence Fonds, VI 13c, Survey re Foreign Water in Manitoba by H.H. McIntyre, September 1946.

6   For drainage district plans, see AM, Manitoba, Executive Council Office, Premier's Office Files, John Bracken, Stuart Garson, Douglas Campbell, GR 43, G 61, file Drainage Districts.

7   Manitoba, Land Drainage Arrangement Commission, *Report,* 33.

8   William A. Waiser, "The Government Explorer in Canada, 1870-1914," in *North American Exploration: A Continent Comprehended,* vol. 3, ed. John Logan Allen (Lincoln: University of Nebraska Press, 1997), 429-30.

9   AM, Manitoba, Drainage Maintenance Boards, Minutes and Office Files, GR 7784, Q-03-26-94, file General, 1941-1946, Memorandum Relating to Drainage District No. 1.

10  AM, Manitoba, Executive Council Office, Clerk's Office, Orders in Council, GR 1530, Order in Council 619, Re Report of the Minister of Public Works on Lands Reclaimed by Drainage, 14 January 1882.

11  Today the area remains wetland and has become the headquarters of Ducks Unlimited.

12  R.T. Riley, "Memoirs" (photocopied manuscript, Legislative Library of Manitoba, n.d.), 61-65.

13  Documents pertaining to such investigations can be found in various sessional papers, including AM, Manitoba, Legislative Assembly, Unpublished Sessional Papers, GR 174, G 8180, file 25; G 8351, files 3 and 11; G 8373, files 3 and 4.

14  On the question of patronage in a Canadian context with a chapter focusing on Manitoba, see Jeffrey Simpson, *Spoils of Power: The Politics of Patronage* (Don Mills, ON: Collins Publishers, 1988). See also Doug Owram, *The Government Generation: Canadian Intellectuals and the State, 1900-1945* (Toronto: University of Toronto Press, 1986), 46-47.

15  Marilyn Baker, *Symbol in Stone: The Art and Politics of a Public Building* (Winnipeg: Hyperion Press, 1986). For more on irregularities in provincial finances, see M.S. Donnelly, *The Government of Manitoba* (Toronto: University of Toronto Press, 1963), 94; and John Kendle, *John Bracken: A Political Biography* (Toronto: University of Toronto Press, 1979),

26-27. On the Manitoba legislative building, see Jonathan F. Vance, *Building Canada: People and Projects that Shaped the Nation* (Toronto: Penguin, 2006), 92-97.

16 Kendle, *John Bracken*, 27.

17 AM, GR 1609, G 8019, file Drainage Commission, Walter L. Coulan to Dr. Hamilton, 29 January 1917.

18 AM, GR 7784, Q 032694, file General, 1941-1946, Memorandum Relating to Drainage District No. 2.

19 AM, GR 1609, G 8019, file Drainage – General, Drainage Commission Notes by Mr. McColl, n.d.

20 AM, GR 1609, G 8019, file Drainage Commission, Report on Land Drainage in the Province of Manitoba by C.G. Elliot, 5 June 1918. See also AM, GR 7784, Q 032694, file General, text of an address delivered at the annual meeting of the Union of Municipal Drainage Districts by F.E. Umphrey, 27 November 1945.

21 C.G. Elliot is discussed as a significant figure in the planning and administration of American drainage projects in Christopher F. Meindl, Derek H. Alderman, and Peter Waylen, "On the Importance of Environmental Claims-Making: The Role of James O. Wright in Promoting the Drainage of Florida's Everglades in the Early Twentieth Century," *Annals of the Association of American Geographers* 92, 4 (2002): 682-701.

22 AM, GR 1609, G 8019, file Drainage Commission, Report on Land Drainage in the Province of Manitoba by C.G. Elliot, 5 June 1918.

23 AM, GR 1530, Order in Council 30724, Report of a Committee of the Executive Council, 17 January 1919.

24 Kendle, *John Bracken*, 26-30.

25 AM, GR 1609, G 8019, file Drainage Commission, Drainage Committee Report, 38-41. See the exchange involving Stewart, Armstrong, and Kennedy.

26 Owram, *The Government Generation*, 45.

27 Owram is building to some extent on earlier scholarship that identified the early decades of the twentieth century as a period of dramatic transformation in Canada and beyond. Robert Craig Brown and Ramsay Cook, *Canada 1896-1921: A Nation Transformed* (Toronto: McClelland and Stewart, 1974); Robert H. Wiebe, *The Search for Order, 1877-1920* (New York: Hill and Wang, 1967).

28 W.L. Morton, *The Progressive Party in Canada* (Toronto: University of Toronto Press, 1950), 33.

29 Owram, *The Government Generation*, 61.

30 J.A. Griffiths, "The History and Organization of Surface Drainage in Manitoba," paper presented before the Winnipeg Branch, Engineering Institute of Canada, 20 March 1952, Sciences and Engineering Library, University of Manitoba, 4-6.

31 Edward Ledohowski, *The Heritage Landscape of the Crow Wing Study Region of Southeastern Manitoba* (Winnipeg: Manitoba Culture, Heritage and Tourism, 2003), 14.

32 AM, Manitoba, Executive Council Office, Premier's Office Files (Campbell Administration), GR 45, N-11-3-16, Notes on Red River Floods with Particular Reference to the Flood of 1950 by R.H. Clark, October 1950, 14.

33 AM, Manitoba, Department of Public Works, Minister's Office Files, GR 1607, G 7987, J.A. Macdonnell to the Minister of Public Works, 30 December 1898.

34 AM, GR 1609, file Drainage Commission, Report on Land Drainage in the Province of Manitoba by C.G. Elliot, 5 June 1918.

35 The use of "highlander" and "lowlander" to describe actors in this debate is also apparent in newspaper reports. For examples, see "Red River Drainage District Extension," *Manitoba*

*Free Press*, 16 February 1922, 4; "Throw Doubts on Accuracy of Drainage Report," *Winnipeg Tribune*, 8 February 1922, 6; "New Drainage District Proposed on Principle of Distributing Cost," *Winnipeg Tribune*, 15 February 1922, 5; "Speakers Urge Distribution of Drainage Costs," *Winnipeg Tribune,* 22 February 1922, 2.

36  W.J. Carlyle, "The Management of Environmental Problems on the Manitoba Escarpment," *Canadian Geographer* 24, 3 (1980): 255-69.

37  AM, Drainage Maintenance Boards Minutes and Office Files, GR 7784, Q 032694, file Drainage – General, Memorandum Related to Drainage District No. 2, n.d.

38  AM, GR 1609, G 8019, file Drainage – General, Drainage Committee Report, L.C. Wilkin Testimony, 110.

39  Manitoba, *Report of the Manitoba Drainage Commission* (Winnipeg: King's Printer, 1921), 11.

40  Ibid.

41  Ibid., 20.

42  The Unpublished Sessional Papers contain an incomplete transcript of the hearings of the Legislative Committee on Drainage about the 1921 report of the Manitoba Drainage Commission. AM, GR 174, G 8374, file 13, Special Committee on Drainage, Drainage Committee Report, Hannesson Testimony, Representative of the Rural Municipality of St. Andrews, 138-39. Another partial transcript is in the files of the Minister of Public Works, AM, GR 1609, G 8019, file Drainage – General, Drainage Committee Report.

43  AM, GR 1609, G 8019, file Drainage Commission, Drainage Committee Report, McCallum Testimony, Representative of the Rural Municipality of Roland, 55. Municipal representatives who spoke in favour of the report came from the Municipalities of Rhineland, Ste. Rose, Glenella, Roland, Morris, Macdonald, Cartier, and Taché. This list draws on both the hearing transcripts and newspaper reports: "Red River Drainage District Extension," *Manitoba Free Press*, 16 February 1922, 4; "Throw Doubts on Accuracy of Drainage Report," *Winnipeg Tribune*, 8 February 1922, 6; "New Drainage District Proposed on Principle of Distributing Cost," *Winnipeg Tribune*, 15 February 1922, 5; "Speakers Urge Distribution of Drainage Costs," *Winnipeg Tribune,* 22 February 1922, 2.

44  AM, GR 174, G 8373, file 13, Drainage Committee Report, Reeve D.F. Stewart Testimony, Representative of the Rural Municipality of Thompson, 33-44. According to the hearing transcripts and media reports, speakers from the Municipalities of Stanley, Thompson, Pembina, Lorne, Grey, Dufferin, Victoria, Ochre River, and South Norfolk disagreed with the report of the Manitoba Drainage Commission.

45  AM, GR 1609, G 8019, file Drainage – General, Drainage Committee Report, Sullivan Testimony, 182.

46  AM, GR 174, G 8374, file 13, H.A. Bowman, Reclamation Branch Chief Engineer, to W.C. McKinnell, Chair, Legislative Committee on Drainage, 16 February 1922.

47  AM, Manitoba, Department of Public Works, Deputy Minister's Office Files, GR 1611, G 8062, file Reclamation Branch, H.A. Bowman to D.G. McKenzie, 12 April 1929.

48  AM, GR 1609, G 8019, file Drainage Commission, Drainage Committee Report, Black Testimony, Representative of the Municipalities of Stanley and Thompson, 21.

49  AM, GR 1609, G 8019, file Drainage Commission, Drainage Committee Report, Goldring Testimony, Representative of the Rural Municipality of Victoria, 49.

50  AM, GR 1609, G 8019, file Drainage Commission, Drainage Committee Report, Reeve D.F. Stewart Testimony, Representative of the Rural Municipality of Thompson, 38.

51  AM, GR 174, G 8374, file 13, Special Committee on Drainage, Drainage Committee Report, J. Rose Testimony, Representative of the Rural Municipality of Ochre River, 126-28.

52  The 1921 census suggests that the lowlands were in fact more agriculturally productive than the highlands, despite drainage problems. Statistics Canada, Table 79, "Farms and Farm Property," 1921 census.

53  AM, GR 1609, G 8019, file Drainage Commission, Drainage Committee Report, Reeve D.F. Stewart Testimony, Representative of the Rural Municipality of Thompson, 35.

54  Mark Fiege, *Irrigated Eden: The Making of an Agricultural Landscape in the American West* (Seattle: University of Washington Press, 1999); see especially Chapter 3, "Dividing Water: Conflict, Cooperation, and Allocation on the Upper Snake River," 81-116.

55  AM, GR 1609, G 8019, file Drainage – General, Drainage Committee Report, L.C. Wilkin Testimony, 112.

56  For examples, see Gerald Friesen, *The Canadian Prairies: A History* (Toronto: University of Toronto Press, 1984); Yossi Katz and John Lehr, *The Last Best West: Essays on the Historical Geography of the Canadian Prairies* (Jerusalem: Magnes Press, 1999); and Thomas Peterson, "Manitoba: Ethnic and Class Politics," in *Canadian Provincial Politics: The Party Systems of the Ten Provinces,* 2nd ed., ed. Martin Robin (Scarborough: Prentice-Hall of Canada, 1978), 61-119.

57  Morton, *Manitoba,* 287-88; Ian MacPherson, *The Co-Operative Movement on the Prairies, 1900-1955* (Ottawa: Canadian Historical Association, 1979), 5.

58  David Laycock, *Populism and Democratic Thought on the Canadian Prairies* (Toronto: University of Toronto Press, 1990), 4-6.

59  Kenneth Michael Sylvester, *The Limits of Rural Capitalism: Family, Culture, and Markets in Montcalm, Manitoba 1870-1940* (Toronto: University of Toronto Press, 2001); see especially Chapter 1, "Shared Origins of Settlement," 12-28; Allan Smith, "The Myth of the Self-Made Man in English Canada, 1850-1914," *Canadian Historical Review* 59, 2 (1978): 189-219; V.C. Fowke, "The Myth of the Self-Sufficient Canadian Pioneer," *Transactions of the Royal Society of Canada* 56, series III (1962): 23-37; Harold Adams Innis, "Foreword," in *Farm Credit in Canada,* by W.T. Easterbrook (Toronto: University of Toronto Press, 1938), v-viii. On the individualist ethos of prairie settlers, see also Nelson Wiseman, "The Pattern of Prairie Politics," in *The Prairie West: Historical Readings,* 2nd ed., ed. R. Douglas Francis and Howard Palmer (Edmonton: University of Alberta Press, 1992), 640-60; and Lyle Dick, *Farmers "Making Good": The Development of Abernethy District, Saskatchewan, 1880-1920* (Calgary: University of Calgary Press, 2008); see especially Chapter 7, "Abernethy's Social Creed,"171-90.

60  AM, GR 1609, G 8019, file Drainage – General, Drainage Committee Report, Reeve Howson Testimony, Representative of the Rural Municipality of Ochre River, 121. Similar arguments were also made in letters to the minister of public works. For example, see AM, GR 1609, G 8019, file Drainage – General, Geo. A. Baker to Premier John Bracken and attached, 26 February 1923.

61  AM, GR 1609, G 8019, file Drainage – General, Drainage Committee Report, Reeve Morten Testimony, Representative of the Rural Municipality of Westbourne, 107.

62  AM, GR 174, G 8374, file 13, Special Committee on Drainage, Black Testimony, Representing the Municipalities of Stanley and Thompson, 90.

63  AM, Manitoba, Drainage Maintenance Boards, Minutes and Office Files, GR 1617, G 5324, G.B. McColl, "Drainage in the Red River Valley in Manitoba," *Canadian Engineer,* 8 September 1917.

64  Fiege makes a similar point in describing the resolution of conflict over the management of stored water along Idaho's Snake River. Fiege, *Irrigated Eden,* 110-12.

65  Morton, *Manitoba,* 386; Donnelly, *The Government of Manitoba,* 122-23.

66  AM, GR 1609, G 8019, file Drainage – General, 1919-1927, McPherson, Minister of Public Works, to Sullivan, Chair of the Drainage Commission, 31 March 1922.

67  AM, GR 1609, G 8019, file Drainage – General, 1919-1927, Sullivan, Chair of the Drainage Commission, to Clubb, Minister of Public Works, 24 July 1923.

68  Peterson, "Manitoba," 77-80.

69  For a discussion of the relationship between science and politics, see Stephen Bocking, *Nature's Experts: Science, Politics, and the Environment* (New Brunswick, NJ: Rutgers University Press, 2004).

70  Warkentin, "Water and Adaptive Strategies in Settling the Canadian West," 72.

71  Ibid.

72  M.A. Lyons, *Report and Recommendations on "Foreign Water" and Maintenance Problems in Drainage Maintenance Districts Constituted under the Land Drainage Arrangement Act, 1935, Province of Manitoba* (Winnipeg: King's Printer, 1950), 7-8.

73  Lenore Eidse, ed., *Furrows in the Valley: Rural Municipality of Morris, 1880-1980* (n.p.: Inter-Collegiate Press, 1980), 35.

74  AM, Manitoba, Executive Council Office, Clerk's Office, Orders in Council, GR 589, Order in Council 50/47, B-15-7-6, 14 January 1947.

75  AM, GR 589, Order in Council 602/35, B-14-8-10, 23 May 1935.

76  AM, GR 1617, G 5324, Survey of Foreign Water in Manitoba by H.H. McIntyre, September 1946.

77  AM, GR 1609, G 8048, file Levies, Attachment to Interdepartmental Memo re Drainage by A.W. Smith, 12 May 1936.

78  Owram, *The Government Generation,* x. For a discussion of this in relation to water control infrastructure, see A.A. den Otter, "Irrigation and Flood Control," in *Building Canada: A History of Public Works,* ed. Norman R. Ball (Toronto: University of Toronto Press, 1988), 158.

79  AM, GR 1609, G 8046, file Union of Municipal Drainage Maintenance Districts, E.F. Willis to M.A. Lyons, 9 May 1945.

80  AM, GR 1609, G 8046, file Special Drainage Survey Conducted by M.A. Lyons, the Lyons Report and Recommendations re "Foreign Water" and Maintenance Problems, n.d.

81  AM, GR 1609, G 8046, file Summary of Drainage Districts, Drainage Districts Committee to W.R. Clubb, 15 February 1934.

82  James Murton, *Creating a Modern Countryside: Liberalism and Land Resettlement in British Columbia* (Vancouver: UBC Press, 2007), 2-3. Sociologist Rod Bantjes has argued that the land and settlement system used on the prairies was in itself thoroughly modern. See Rod Bantjes, *Improved Earth: Prairie Space as Modern Artefact* (Toronto: University of Toronto Press, 2005).

83  AM, GR 1609, G 8019, file Drainage – General, Drainage Committee Report, Sullivan Testimony, 161.

## CHAPTER 4: INTERNATIONAL BIOREGIONS AND LOCAL MOMENTUM

1  On the concept of efficiency and its evolution, see Samuel P. Hays, *Conservation and the Gospel of Efficiency: The Progressive Conservation Movement, 1890-1920* (Cambridge, MA: Harvard University Press, 1959), 122-27; Samuel P. Hays, *Beauty, Health, and Permanence: Environmental Politics in the United States, 1955-1985* (New York: Cambridge University

Press, 1987), 208-13; and Hugh S. Gorman, "Efficiency, Environmental Quality, and Oil Field Brines: The Success and Failure of Pollution Control by Self-Regulation," *Business History Review* 73, 4 (1999): 601-40.

2  W. Edgar Watt, "The National Flood Damage Reduction Program: 1976-1995," *Canadian Water Resources Journal* 20, 4 (1995): 237-48.

3  N.F. Dreisziger, "The International Joint Commission of the United States and Canada, 1895-1920: A Study in Canadian-American Relations" (PhD diss., University of Toronto, 1974), 212-14.

4  William Carlyle, "Water in the Red River Valley of the North," *Geographical Review* 74 (1984): 331-58; Warren Upham, *The Glacial Lake Agassiz,* Monographs of the United States Geological Survey 25 (Washington, DC: United States Geological Survey, 1895), 14-15. For a discussion of the north-south factor in North American history, see Elliot West, "Against the Grain: State-Making, Cultures, and Geography in the American West," in *One West, Two Myths: A Comparative Reader,* ed. C.L. Higham and Robert Thacker (Calgary: University of Calgary Press, 2004), 1-21. An influential Canadian advocate of north-south integration was Goldwin Smith, *Canada and the Canadian Question* (Toronto: Hunter, Rose, 1891). See also Graeme Wynn, "Forging a Canadian Nation," in *North America: The Historical Geography of a Changing Continent,* ed. Robert D. Mitchell and Paul A. Groves (Totowa, NJ: Rowman and Littlefield, 1987), 373-409.

5  Kim Richard Nossal, "Institutionalization and the Pacific Settlement of Interstate Conflict: The Case of Canada and the International Joint Commission," *Journal of Canadian Studies* 18, 4 (1983-84): 79.

6  AM, Manitoba, Executive Council Office, Premier's Office Files (Campbell Administration), GR 45, N 11316, Canada, Department of Resources and Development, *Report on Investigations into Measures for the Reduction of the Flood Hazard in the Greater Winnipeg Area* (Ottawa: Water Resources Division, Red River Basin Investigation, 1953), Appendix A, "Geography and Development," 2.

7  LAC, Canada, Water Resources Branch Records, RG 89, vol. 227, file 2692, Louis Coste to E.F.E. Roy, 28 February 1893.

8  Edward Ledohowski, *The Heritage Landscape of the Crow Wing Study Region of Southeastern Manitoba* (Winnipeg: Manitoba Culture, Heritage and Tourism, 2003), 12-13.

9  Canada, *Sessional Papers,* 1896, Department of the Interior Report, 12-13.

10  John Lehr, "The Process and Pattern of Ukrainian Rural Settlement in Western Canada, 1891-1914" (PhD diss., University of Manitoba, 1978); Nickolaus Waggenhoffer, "Some Socio-Economic Dynamics in South-Eastern Manitoba, with Particular Reference to the Farming Communities within the Local Government Districts of Stuartburn and Piney" (MA thesis, University of Manitoba, 1972); John Lehr, "'Shattered Fragments': Community Formation on the Ukrainian Frontier of Settlement, Stuartburn, Manitoba, 1896-1921," *Prairie Forum* 28, 2 (2003): 219-34; John Lehr, "The Peculiar People: Ukrainian Settlement of Marginal Lands in Southeastern Manitoba," in *Building beyond the Homestead: Rural History on the Prairies,* ed. Ian MacPherson and David Jones (Calgary: University of Calgary Press, 1985), 29-46; Gerald Friesen, "Three Notes on the History of Rural Manitoba," in *River Road: Essays on Manitoba and Prairie History* (Winnipeg: University of Manitoba Press, 1996), 201; Yossi Katz and John Lehr, *The Last Best West: Essays on the Historical Geography of the Canadian Prairies* (Jerusalem: Magnes Press, 1999), especially Chapter 3, "The Creation of Ukrainian Block Settlement," 62-75, and Chapter 4, "The Social Structure of Ukrainian Settlement," 76-98.

11  LAC, Canada, Department of Public Works, RG 11, vol. 4370, file 7438-2-A, Field Supervisor, Soldier Settlement Board, to G.C. Cummings, 20 October 1927.
12  Lehr, "The Peculiar People," 35.
13  AM, Manitoba, Department of Public Works, Minister's Office Files, GR 1607, G 7980, item 874, Report on Drainage Municipality of Franklin by John Molloy, 5 August 1896.
14  Ibid.
15  AM, GR 1607, G 7988, item 411, L. Ross to Robert Watson, 3 April 1899.
16  Gordon B. Dodds, "The Stream-Flow Controversy: A Conservation Turning Point," *Journal of American History* 56, 1 (1969): 59-69; Hays, *Conservation and the Gospel of Efficiency*, 203-5; Nancy Langston, *Forest Dreams, Forest Nightmares: The Paradox of Old Growth in the Inland West* (Seattle: University of Washington Press, 1995), 141-48; Luna B. Leopold and Thomas Maddock Jr., *The Flood Control Controversy: Big Dams, Little Dams, and Land Management* (New York: Ronald Press Company, 1954); James L. Wescoat Jr., "'Watersheds' in Regional Planning," in *The American Planning Tradition: Culture and Policy*, ed. Robert Fishman (Baltimore: Johns Hopkins University Press, 2000), 147-71.
17  AM, Manitoba, Legislative Assembly, Unpublished Sessional Papers, GR 174, G 8262, Navigation of Red River, 1908, and G 8375, Return to an Order of the House (No. 44) re Flood Control of Red River, 1922.
18  LAC, Canada, International Joint Commission Records, RG 51, vol. 1, "International Joint Commission Waterways Problems," by L.J.B., 17 September 1923. The author states that the Roseau River matter is an example of a situation that could develop along the Red River but at a smaller scale.
19  LAC, RG 11, vol. 4369, file 7438-1-C, Prudhomme to Cory, 18 June 1928; LAC, RG 11, vol. 4369, file 7438-1-C, Bracken to King, 14 June 1928; LAC, RG 89, vol. 580, file 963, Historical Summary of the Roseau River Flood Investigations Compiled by H.R. Cram, December 1925, Appendix 7, Charles Stewart to O.D. Skelton, 19 March 1927.
20  Herbert A. Hard, *Report to the Governor of North Dakota on Flood Control, 1919-1920* (Grand Forks: Normanden Publishing Company, n.d.).
21  LAC, RG 89, vol. 227, file 2692, Memo re Flood on Roseau River to W.W. Cory from R.J.B., 5 September 1919. A handwritten note signed with indecipherable initials has been added to this typed memo: "A matter for the province to deal with – until at all events it passes a Reclamation Act similar to Sask. & Alberta." See also LAC, RG 89, vol. 227, file 2692, Drake to J.G. Sullivan, Chair, Manitoba Drainage Commission, 16 September 1919; and LAC, RG 89, vol. 580, file 963, History of Roseau River Floods, 21 December 1925.
22  LAC, RG 89, vol. 580, file 963, Historical Summary of the Roseau River Flood Investigations Compiled by H.R. Cram, December 1925, Appendix 7, F.E. Drake to J.M. Mysyk, 23 July 1915; LAC, RG 11, vol. 4370, file 7438-2-A, Robert Semple to Robert Rogers, 22 October 1930; LAC, RG 89, vol. 227, file 2692, Memorandum for File, 27 March 1927.
23  LAC, RG 89, vol. 580, file 963, Ralph J. Burley to W.W. Cory, 5 September 1919.
24  Geoff Cunfer and Dennis Guse, "Ditches," in *Draining the Great Oasis: An Environmental History of Murray County, Minnesota*, ed. Anthony J. Amato, Janet Timmerman, and Joseph A. Amato (Marshall, MN: Crossings Press, 2001), 145.
25  LAC, RG 89, vol. 595, file 1288A, Statement Relative to Roseau River and the Pending International Questions Prepared by United States Geological Survey, n.d.
26  LAC, RG 89, vol. 227, file 2692, J.M. Mysyk to Department of the Interior, 5 August 1919.

27  LAC, RG 89, vol. 227, file 2692; LAC, RG 89, vol. 580, file 963, History of Roseau River
    Floods, 21 December 1925.

28  LAC, RG 11, vol. 4370, file 7438-2-A, Field Supervisor, Soldier Settlement Board, to G.C.
    Cummings, 20 October 1927.

29  LAC, RG 11, vol. 4369, file 7438-1-A, F.G. Goodspeed, District Engineer, to K.M. Cameron,
    Chief Engineer, 21 April 1926.

30  LAC, RG 89, vol. 596, file 1312, copy of a letter sent from Premier Bracken to O.D. Skelton,
    14 November 1928, included in the appendix to a brief filed for the Department of Public
    Works of Canada by R. de B. Corriveau. See also LAC, RG 11, vol. 4369, file 7438-1-A,
    Goodspeed to Cameron, 21 April 1927; and LAC, RG 11, vol. 4369, file 7438-1-B. The latter
    file contains a flurry of letters from Prudhomme to various Manitoba and dominion of-
    ficials on the matter.

31  Robert Michael William Graham, "The Surface Waters of Winnipeg: Rivers, Streams,
    Ponds, and Wetlands, 1874-1984: The Cyclical History of Urban Land Drainage" (MLA
    Practicum, University of Manitoba, 1984). It is evident that urban land drainage proceeded
    at a pace at least equal to that of rural land drainage. With regard to a 181.99 square kilo-
    metre area including what is now Winnipeg, estimates put the wetland area in 1800 at
    134.01 square kilometres. In the late twentieth century, the wetland area was estimated at
    3.96 square kilometres. See Manitoba, Water Resources Branch, *Papers of the Third Annual
    Western Provincial Conference: Rationalization of Water and Soil Research and Management*
    (Winnipeg: Queen's Printer, 1985), 423.

32  LAC, RG 11, vol. 4369, file 7438-1-A, J.G. Sullivan, "The Red River Flood Problem,"
    *Engineering News-Record,* 20 April 1922; AM, GR 1611, G 8064, file Reclamation Branch,
    M. Peterson to A. McGillivray, 24 July 1930.

33  LAC, RG 11, vol. 4369, file 7438-1-A, J. Prudhomme to Dominion Government, 5 March
    1927. In this document, Prudhomme indicates that he had learned of the reference to the
    IJC from a report in a local newspaper.

34  LAC, RG 89, vol. 660, file 2339, International Joint Commission Hearings re the Roseau
    River Reference, 289-90.

35  Nossal, "Institutionalization and the Pacific Settlement of Interstate Conflict," 79; emphasis
    in original.

36  LAC, RG 89, vol. 69, file 64, vol. 1, J.T. Johnston to R.C. Williams, 30 April 1928.

37  Nossal, "Institutionalization and the Pacific Settlement of Interstate Conflict," 75-87.

38  For more on the working norms of the International Joint Commission, see William R.
    Willoughby, "Expectations and Experience, 1909-1979," in *The International Joint Com-
    mission Seventy Years On,* ed. Robert Spencer, John Kirton, and Kim Richard Nossal
    (Toronto: T.H. Best Printing, 1981), 24-42; L.M. Bloomfield, *Boundary Waters Problems of
    Canada and the United States* (Toronto: Carswell, 1958).

39  LAC, RG 89, vol. 660, file 2339, International Joint Commission Hearings re the Roseau
    River Reference, 82-83.

40  The files of the Manitoba Office of Ducks Unlimited (Canada) in Brandon contain a copy
    of a letter from R.K. Brace, Populations Biologist, Environment Canada, Environmental
    Management, Canadian Wildlife Service, to Crawford Jenkins, Watershed Conservation
    Districts, Water Resources, 14 April 1977. Brace says that he continues "to be amazed that
    engineers from Water Resources believe that upstream flooding has no effect on downstream
    flooding." Discussions of upstream/downstream linkages (particularly in respect to the
    connection between drainage and flooding) in Manitoba include AM, GR 45, N 11316,
    Notes on Red River Floods with Particular Reference to the Flood of 1950, by R.H. Clark,

October 1950, 15; K. Juliano and S.P. Simonovic, "The Impact of Wetlands on Flood Control in the Red River Valley of Manitoba" (Natural Resources Institute, University of Manitoba, 1999), 10-11, 62-63; W.F. Rannie, "Manitoba Flood Protection and Control Strategies," *Proceedings of the North Dakota Academy of Science* 53 (1999): 38-43; and Gene Krenz and Jay Leitch, *A River Runs North: Managing an International River* (n.p.: Red River Water Resources Council, 1993), 51-56. For a succinct general discussion, see William B. Meyer, "When Dismal Swamps Became Priceless Wetlands," *American Heritage* 45, 3 (1994): 108.

41   AM, GR 174, G 8364, file 13, 80, 169-70; GR 1617, G 5324, Drainage Committee Report, 15 February 1922.

42   AM, Manitoba, Department of Public Works, Minister's Office Files, GR 1609, G 8019, file Drainage – General, Speech Made by Mr. Clubb to House on Drainage, 10 May 1923.

43   For further information on IJC involvement in the management of Roseau River, see Bloomfield, *Boundary Waters Problems of Canada and the United States,* 139-40.

44   LAC, RG 11, vol. 4369, file 7438-1-C, Goodspeed to Cameron, 28 August 1929.

45   Krenz and Leitch, *A River Runs North,* 1.

46   Upham, *The Glacial Lake Agassiz,* 587. Other sources have estimated the size of the Big Grass Marsh as 40,000 acres. W.G. Leitch, *Ducks and Men: Forty Years of Co-Operation in Conservation* (Winnipeg: Ducks Unlimited Canada, 1978), 39.

47   LAC, Canada, Department of the Interior, RG 15, D-II-I, reel T-12178, vol. 230, file 223, two petitions from Josepth Little and others asking for an appropriation to drain the "Big Grass" into the "White Mud River," 1873.

48   AM, GR 174, G 8128, C.P. Brown to unnamed, 6 April 1882.

49   AM, GR 174, G 8128, file 6, Return to an Order re Transactions of Reclaimed Lands, 1886.

50   AM, GR 1607, G 7959, G.B. Bemister to C.P. Brown, 3 May 1880.

51   AM, GR 1607, G 7959, Agreement between the Manitoba Government and Elijah Griffith. See also Manitoba, Department of Public Works, *Annual Report,* 1880.

52   AM, GR 174, G 8128, Return to an Order re Transactions of Reclaimed Lands, 1886.

53   AM, GR 1607, G 7959, Agreement between the Manitoba Government and the Manitoba Drainage Company, n.d.

54   AM, GR 1607, G 7959, Contract with the Manitoba Drainage Company, 10 June 1880.

55   AM, GR 1670, G 7389, Westbourne Municipal Council Minute Book, 24 June 1879.

56   AM, GR 1670, G 7379, Westbourne Municipal Council Minute Book, 10 May 1899.

57   AM, GR 174, G 8221, Annual Report of the Department of Public Works, 1902.

58   Known locally as the Kinisota (or Kinnesota or Langruth) Ridge, the landform was part of a glacial feature known as the Burnside Beach, which marks one of the former shorelines of Lake Agassiz. Upham, *The Glacial Lake Agassiz,* 467; John Warkentin, "Human History of the Glacial Lake Agassiz Region in the 19th Century," in *Life, Land, and Water: Proceedings from a Conference on the Environmental Studies of the Glacial Lake Agassiz Region,* ed. William Mayer-Oakes (Winnipeg: University of Manitoba Press, 1967), 327. For more on the historical geography of the area, see William Carlyle, "The Relationship between Settlement and the Physical Environment in Part of the West Lake Area of Manitoba from 1878 to 1963" (MA thesis, University of Manitoba, 1965).

59   AM, GR 174, G 8271, Annual Report of the Department of Public Works, 1898.

60   AM, GR 174, G 8271, Annual Report of the Department of Public Works, 1909.

61   Manitoba, Land Drainage Arrangement Commission, *Report of the Land Drainage Arrangement Commission Respecting Municipalities Containing Land Subject to Review under "The Land Drainage Act"* (Winnipeg: King's Printer, 1936), 32.

62  AM, GR 174, G 8271, Annual Report of the Department of Public Works, 1909.

63  Manitoba, Land Drainage Arrangement Commission, *Report,* 33.

64  *Gladstone Age,* 30 October 1913, quoted in *Return to Big Grass,* 2nd ed., ed. Richard Wentz (n.p.: Ducks Unlimited, 1986), 8.

65  Office Files of Ducks Unlimited, Manitoba Branch, Brandon (hereafter DU Office), "A Brief History of Big Grass Marsh," by B.W. Cartwright.

66  Manitoba, Committee on Utilization of Public Lands in the Province of Manitoba, *Report Submitted to the Minister of Agriculture: An Overview* (Winnipeg: Manitoba Department of Agriculture, 1934), 8-18. The report argues that Drainage Districts No. 8 and No. 19 (those in the vicinity of Big Grass Marsh) should not have been created.

67  AM, Manitoba, Drainage Maintenance Boards, Minutes and Office Files, GR 7784, Memorandum for W.R. Clubb, 24 January 1935.

68  Robert Murchie, *The Unused Lands of Manitoba: Report of Survey* (Winnipeg: Department of Agriculture and Immigration, 1926), 129-31.

69  Richard Wentz, ed., *Return to Big Grass* (n.p.: Ducks Unlimited, 1986), 9.

70  DU Office, Letter to the Editor of the *Gladstone Age,* 9 August 1948.

71  DU Office, "A Brief History of Big Grass Marsh," by B.W. Cartwright. See also Leitch, *Ducks and Men,* 39.

72  DU Office, "A Brief History of Big Grass Marsh," by B.W. Cartwright.

73  DU Office, Letter to the Editor of the *Gladstone Age,* 9 August 1948.

74  AM, GR 174, G 8340, file 1, John Bracken to R.B. Bennett, Prime Minister of Canada, 19 August 1931.

75  AM, GR 174, G 8257, Annual Report of the Department of Mines and Natural Resources, 1937.

76  DU Office, "A Brief History of Big Grass Marsh," by B.W. Cartwright.

77  Joseph James Hargrave, *Red River* (Montreal: John Lovell, 1871), 173.

78  On the flyway concept, see Robert M. Wilson, *Seeking Refuge: Birds and Landscapes of the Pacific Flyway* (Seattle: University of Washington Press, 2010). Chapters 1 and 3 are of particular interest.

79  George Colpitts, *Game in the Garden: A Human History of Wildlife in Western Canada to 1940* (Vancouver: UBC Press, 2002), 63-64, 75-102.

80  Kurkpatrick Dorsey, "Scientists, Citizens, and Statesmen: US-Canadian Wildlife Protection Treaties in the Progressive Era," *Diplomatic History* 19, 3 (1995): 407-29; Kurkpatrick Dorsey, *The Dawn of Conservation Diplomacy: U.S.-Canadian Wildlife Protection Treaties in the Progressive Era* (Seattle: University of Washington Press, 1998), 165-215.

81  AM, GR 174, G 8360, Annual Report of the Department of Mines and Natural Resources, 1935.

82  AM, GR 174, G 8251, Annual Report of the Department of Mines and Natural Resources, 1933.

83  Ibid.

84  Leitch, *Ducks and Men,* 14. See also Wentz, ed., *Return to Big Grass,* viii.

85  Tina Loo, *States of Nature: Conserving Canada's Wildlife in the Twentieth Century* (Vancouver: UBC Press, 2006), 287.

86  Wentz, ed., *Return to Big Grass,* 4-5.

87  Leitch, *Ducks and Men,* 18.

88  Report reproduced in ibid., 19.

89 AM, Manitoba, Executive Council Office, Premier's Office Files, John Bracken, Stuart Garson, Douglas Campbell, GR 43, G 64, file Ducks Unlimited, press release prepared by T.C. Main in conjunction with the 5 April 1941 opening of the Bracken Dam.

90 Leitch, *Ducks and Men*, 21.

91 Ibid., 27.

92 Report reproduced in ibid., 19.

93 Ibid., 27.

94 Ibid., 199.

95 AM, Manitoba, Department of Mines and Natural Resources, Deputy Minister's Files, GR 1600, G 4530, file Game Branch, Ducks Unlimited, E.S. Russenholt to J.S. McDiarmid, with attachments, 8 February 1939.

96 Ibid.

97 Ibid.

98 DU Office, "The Lesson of Big Grass Marsh," by William G. Leitch, *Ducks Unlimited Magazine,* spring 1968.

99 Manitoba, Land Drainage Arrangement Commission, *Report,* 63.

100 DU Office, E.B. Pitblado to Gurney Evans, 2 December 1958. See Figure 14 for locations of these municipalities.

101 Leitch, *Ducks and Men,* 41.

102 AM, GR 1600, G 4530, file Ducks Unlimited, C.H. Attwood to E.B. Pitblado, 7 December 1937. See also Wentz, ed., *Return to Big Grass,* 24.

103 Leitch, *Ducks and Men,* 40.

104 AM, GR 43, G 72, Minutes of Ducks Unlimited Annual Meeting, Winnipeg, 1942.

105 For a discussion of some other disagreements, see Thomas R. Dunlap, *Saving America's Wildlife: Ecology and the American Mind, 1850-1990* (Princeton: Princeton University Press, 1988), 48-61, 84-97.

106 AM, H. Albert Hochbaum Fonds, Series 4, P7179/24, file Outward/Inward Correspondence with Ducks Unlimited, item 7, Ira N. Gabrielson, Director of Fish and Wildlife Service, United States Department of the Interior, to T.C. Main, General Manager, DU (Canada), 23 May 1941.

107 AM, H. Albert Hochbaum Fonds, Series 4, P7177/14, item 64, file Outward/Inward Correspondence with Aldo S. Leopold, Alfred Hochbaum to Aldo Leopold, 7 September 1945.

108 AM, H. Albert Hochbaum Fonds, Series 4, P7179/24, file Outward/Inward Correspondence with Ducks Unlimited, item 62, Aldo Leopold to M.W. Smith, Ducks Unlimited (US), 8 July 1947.

109 AM, GR 1600, G 4529, file 11.2.1, Game Branch, Ducks and Geese, General, 1932-1951, Dane MacCarthy to J.S. McDiarmid, 1 September 1947.

110 AM, GR 1600, G 4529, file 11.2.1, Game Branch, Ducks and Geese, General, 1932-1951, Robert Hawkins, MLA, to Minister of Mines and Natural Resources J.S. McDiarmid, 16 September 1946.

111 Ibid. For an analysis of farmers' reactions to waterfowl that ate crops along the Pacific flyway, see Robert Michael Wilson, "Seeking Refuge: Making Space for Migratory Waterfowl and Wetlands along the Pacific Flyway" (PhD diss., University of British Columbia, 2003), 159-70.

112 AM, Edgar S. Russenholt Fonds, P 2829/16, Drought and Rehabilitation "A Personal Experience."

113 LAC, RG 89, file 2339, International Joint Commission Hearings re the Roseau River Reference, 6-7 June 1929.

114 AM, GR 174, G 8340, file 1, Annual Report of the Department of Mines and Natural Resources, 1931.

115 Janet Foster, *Working for Wildlife: The Beginning of Preservation in Canada* (Toronto: University of Toronto Press, 1978), 12-15.

116 Gerald Friesen, *The West: Regional Ambitions, National Debates, Global Age* (Toronto: Penguin/ McGill Institute, 1999), 115-18.

117 J.H. Ellis, *Manitoba Agriculture and Prairie Farm Rehabilitation Activities* (Winnipeg: Department of Agriculture, 1944), 14.

118 Deborah Fitzgerald, *Every Farm a Factory: The Industrial Ideal in American Agriculture* (New Haven: Yale University Press, 2003), 3-9.

119 On the relationship between wetland conservation and drainage in Ontario, see John C. McLaughlin, "Progress, Politics, and the Role of Conservation: Wetland Drainage in Ontario" (PhD diss., Queen's University, 1995), 251-53.

120 Thomas P. Hughes, "Technological Momentum," in *Does Technology Drive History? The Dilemma of Technological Determinism*, ed. Merritt Roe Smith and Leo Marx (Cambridge, MA: MIT Press, 1994), 101-13. Also helpful here is Martin Melosi's work on path dependence in *The Sanitary City: Urban Infrastructure in America from Colonial Times to the Present* (Baltimore: Johns Hopkins University Press, 2000), 10-12.

121 AM, Manitoba, Drainage Maintenance Boards, Minutes and Office Files, GR 1617, G 5324, Memorandum re the Drainage Problem and a Suggested Solution, 14 April 1947.

## CHAPTER 5: PERMANENCE, MAINTENANCE, AND CHANGE

1 William R. Newton, "Watershed Conservation Districts in Manitoba: The Aims, Functions, and Objectives," paper presented at the 89th Annual Congress of the Engineering Institute of Canada, Winnipeg, 1 October 1975, Conservation and Environment Library, Government of Manitoba.

2 Barry Potyondi, *Selkirk: The First Hundred Years* (Winnipeg: Josten's/National School Services, 1981), 11.

3 Sandford Fleming, *Reports on Bridging Red River* (Ottawa: n.p., 1879). Available as Canadian Institute for Historical Microreproductions No. 06108. On the railway matter, see also Alan Artibise, *Winnipeg: A Social History of Urban Growth, 1874-1914* (Montreal: McGill-Queen's University Press, 1975), 61-76.

4 J.M. Bumsted, *Floods of the Centuries: A History of Flood Disasters in the Red River Valley, 1776-1997* (Winnipeg: Great Plains Publications, 1997), 7-10; Doug Owram, *Promise of Eden: The Canadian Expansionist Movement and the Idea of the West, 1856-1900* (Toronto: University of Toronto Press, 1980), 101-24.

5 Bumsted, *Floods of the Centuries*, 9-10, 26.

6 *The Nor'wester*, 1 June 1861, quoted in A.A. den Otter, "Irrigation and Flood Control," in *Building Canada: A History of Public Works*, ed. Norman R. Ball (Toronto: University of Toronto Press, 1988), 158; Samuel Matheson, "Floods at Red River," in *A Thousand Miles of Prairie: The Manitoba Historical Society and the History of Western Canada*, ed. Jim Blanchard (Winnipeg: University of Manitoba Press, 2002), 243-44; AM, MG 7, reel A 83, Records of the Church Missionary Society, David Anderson, Notes of the Flood at Red River by the Bishop of Rupert's Land, 1852.

7   Matheson, "Floods at Red River," 243-44.

8   Bumsted, *Floods of the Centuries,* 24-26.

9   *Journals of the Manitoba Legislative Assembly,* 2nd session, 1st Legislature, 1872, 36. The matter is noted in Edgar Stanford Russenholt, *Heart of the Continent, Being the History of Assiniboia* (Winnipeg: Macfarlane Communications Services, 1968), 151.

10  Scott St. George and Erik Nielsen, "Paleoflood Records for the Red River, Manitoba, Canada, Derived from Anatomical Tree-Ring Signatures," *The Holocene* 13, 4 (2003): 547-55. See also Gregory R. Brooks et al., *Geoscientific Insights into Red River Flood Hazards in Manitoba: The Final Report of the Red River Flood Project* (Ottawa: Natural Resources Canada, 2003).

11  J.M. Bumsted, "The Manitoba Royal Commission on Flood Cost Benefit and the Origins of Cost Benefit Analysis in Canada," *American Review of Canadian Studies* 32, 1 (2002): 98.

12  AM, Manitoba, Executive Council Office, Premier's Office Files (Campbell Administration), GR 45, N-11-3-16, Notes on Red River Floods with Particular Reference to the Flood of 1950, by R.H. Clark, October 1950.

13  Hugh Prince, *Wetlands of the American Midwest: A Historical Geography of Changing Attitudes* (Chicago: University of Chicago Press, 1997), 1-25; John Thompson, *Wetlands Drainage, River Modification, and Sectoral Conflict in the Lower Illinois Valley,* 1890-1930 (Carbondale: Southern Illinois University Press, 2002), 16-20; Ann Vileisis, *Discovering the Unknown Landscape: A History of America's Wetlands* (Washington, DC: Island Press, 1997), 29-50.

14  AM, Manitoba, Department of Natural Resources, Miscellaneous Land Files, GR 7721, reel M 1690, file 5, Extract of a Report of a Committee of the Privy Council, 8 April 1880. See also the *Journals of the Manitoba Legislative Assembly,* 3rd Session, 4th Legislature, 44 Victoria 1881, Appendix B.

15  Manitoba, *Annual Report of the Manitoba Department of Public Works,* 1880, Extract from J. Norquay to A. Campbell, Receiver General, Ottawa, 24 March 1879.

16  *Journals of the Manitoba Legislative Assembly,* 3rd Session, 4th Legislature, 1881, Appendix B.

17  Leslie Hewes and Phillip E. Frandson, "Occupying the Wet Prairie: The Role of Artificial Drainage in Story County, Iowa," *Annals of the Association of American Geographers* 42, 1 (1952): 40.

18  For a similar argument, see Mark Fiege, *Irrigated Eden: The Making of an Agricultural Landscape in the American West* (Seattle: University of Washington Press, 1999), 40-41.

19  AM, Manitoba, Legislative Assembly, Unpublished Sessional Papers, GR 174, G 8218, Department of Agriculture and Immigration Annual Report, 1901.

20  AM, GR 174, G 8374, file 13, H.A. Bowman to W.C. McKinnell, 16 February 1922.

21  AM, GR 174, G 8379, file 2, Department of Public Works Annual Report, 1927.

22  J.H. Ellis, *The Soils of Manitoba* (Winnipeg: Manitoba Economic Survey Board, 1938), 31.

23  AM, GR 174, G 8379, file 1, Department of Public Works Annual Report, 1926.

24  AM, GR 174, G 8379, file 4, Department of Public Works Annual Report, 1930.

25  AM, GR 174, G 8367, file 2, Department of Agriculture and Immigration Annual Report, 1922.

26  H.E. Woods, "Keeping Up with 2,4-D," *Country Guide,* February 1948, 5.

27  AM, Manitoba, Department of Public Works, Minister's Office Files, GR 1609, G 8019, letter (unsigned and unaddressed), 9 February 1924. On the spread of weeds on the Canadian Prairies more generally, see Clinton L. Evans, *The War on Weeds in the Prairie West: An Environmental History* (Calgary: University of Calgary Press, 2002). For a similar American story, see Mark Fiege, "The Weedy West: Mobile Nature, Boundaries, and Common Space in the Montana Landscape," *Western Historical Quarterly* 35, 1 (2005): 22-47.

28   AM, Manitoba, Drainage Maintenance Boards, Minutes and Office Files, GR 1617, G 5324, G.B. McColl, "Drainage in the Red River Valley in Manitoba," *Canadian Engineer,* 8 September 1917.

29   Manitoba, Land Drainage Arrangement Commission, *Report of the Land Drainage Arrangement Commission Respecting Municipalities Containing Land Subject to Review under "The Land Drainage Act"* (Winnipeg: King's Printer, 1936), 7-10; AM, GR 1609, G 8019, file Drainage Commission, Report on Land Drainage in the Province of Manitoba, by C.G. Elliot, 5 June 1918.

30   AM, Manitoba, Department of Public Works, Deputy Minister's Office Files, GR 1610, reel M 932, 193, C.H. Dancer to J.G. Harvey, 4 May 1909.

31   Manitoba, *Report of the Manitoba Drainage Commission* (Winnipeg: King's Printer, 1921), 9.

32   Ibid. For further evidence that Sullivan understood that a key aspect of the problem was the tension between the municipality and the drainage district, see AM, GR 1609, G 8019, file Drainage – General, Drainage Committee Report; note especially 131-58.

33   AM, Manitoba, Executive Council Office, Premier's Office Files, John Bracken, Stuart Garson, Douglas Campbell, GR 43, G 61, file Drainage Districts, Drainage Districts Committee to W.R. Clubb, 15 February 1934.

34   AM, Manitoba, Drainage Maintenance Boards, Minutes and Office Files, GR 7784, Q 032694, file Miscellaneous Reports Drainage Districts 1934, C.L. Stoney et al. to W.R. Clubb, 15 February 1934.

35   AM, GR 43, G 61, file Drainage Districts, Report Submitted to W.R. Clubb, Minister of Public Works, 15 February 1934. For more on this, see AM, GR 174, G 8233, file 3, Select Standing Committee on Agriculture and Immigration – Reports, 1933.

36   Manitoba, Land Drainage Arrangement Commission, *Report,* 16.

37   AM, GR 1609, G 8048, file Levies, Memorandum for W.R. Clubb re Recommendations to Council Respecting Drainage Settlement, n.d.

38   Ibid.

39   AM, GR 7784, Q 032694, file Letters etc. Drainage Maintenance Boards, F.E. Umphrey to W.R. Clubb, 4 March 1938.

40   AM, GR 1617, G 5324, Address Delivered by F.E. Umphrey at the Union of Municipal Drainage Districts, 27 November 1945.

41   Ibid.

42   AM, GR 1609, G 8046, file Drainage Maintenance Boards Overexpenditure 1944, 1945. Union of Municipal Drainage Maintenance Boards, President's Address at 3rd Annual Convention, 9 January 1947. AM, GR 7784, Q 032694, file General, M.A. Lyons to F.E. Umphrey, 10 December 1945.

43   AM, GR 7784, Q 032694, file General, M.A. Lyons to F.E. Umphrey, 10 December 1945.

44   AM, Executive Council Office, Clerk's Office, Orders in Council, GR 589, Order in Council 50/47, 14 January 1947. Lyons's appointment was made retroactive for some months, which suggests that Lyons had already been working on the matter.

45   AM, GR 1609, G 8046, file Special Drainage Survey Conducted by M.A. Lyons, M.A. Lyons to E.F. Willis, n.d.

46   J.A. Griffiths, "The History and Organization of Surface Drainage in Manitoba," paper presented before the Winnipeg Branch, Engineering Institute of Canada, 20 March 1952, Sciences and Engineering Library, University of Manitoba, 4-7.

47   AM, GR 1609, G 8046, file Special Drainage Survey Conducted by M.A. Lyons, J.A. Griffiths to G. Collins, 11 June 1951.

48 AM, GR 1609, G 8046, file Union of Municipal Drainage Maintenance Districts, text of an address delivered by F.E. Umphrey at the annual meeting of the Union of Municipal Drainage Districts, 27 November 1945. Capitalization in original.
49 Timothy Bruce Krywulak, "An Archaeology of Keynesianism: The Macro-Political Foundations of the Modern Welfare State in Canada, 1896-1948" (PhD diss., Carleton University, 2005); see especially Chapter 6, "The Second National Policy: The Keynesian Revolution Comes to Canada, 1939-1948." See also Robert Malcolm Campbell, *Grand Illusions: The Politics of the Keynesian Experience in Canada, 1945-1975* (Peterborough: Broad-view Press, 1987); in particular, see Section 2, "The Keynesian Synthesis in Canada," 31-68.
50 M.A. Lyons, *Report and Recommendations on "Foreign Water" and Maintenance Problems in Drainage Maintenance Districts Constituted under the Land Drainage Arrangement Act, 1935, Province of Manitoba* (Winnipeg: King's Printer, 1950), 1.
51 AM, GR 1609, G 8046, file Special Drainage Survey Conducted by M.A. Lyons, Outline of Work Required in Connection with Investigation of Drainage Districts, by M.A. Lyons, n.d. See also in this file the Lyons Report and Recommendations re "Foreign Water" and Maintenance Problems, n.d., as well as M.A. Lyons to E.F. Willis, 9 September 1947.
52 AM, GR 589, Order in Council 50/47, 14 January 1947.
53 AM, GR 1609, G 8046, file Special Drainage Survey Conducted by M.A. Lyons, Deputy Minister to E.F. Willis, 17 September 1946.
54 AM, GR 1609, G 8046, file Special Drainage Survey Conducted by M.A. Lyons, M.A. Lyons to E.F. Willis, Minister of Public Works, 31 March 1948. See also Lyons, *Report and Recommendations*, 16.
55 Barry Potyondi, *In Palliser's Triangle: Living in the Grasslands, 1850-1930* (Saskatoon: Purich Publishing, 1995), 91-93.
56 Ellis, *The Soils of Manitoba*, 39-40.
57 J.H. Ellis, *Manitoba Agriculture and Prairie Farm Rehabilitation Activities* (Winnipeg: Department of Agriculture, 1944), 7-8.
58 Newton, *Watershed Conservation Districts in Manitoba*, 8.
59 J.C. MacPherson, *A Brief History of the Whitemud Watershed Committee* (Winnipeg: Department of Mines, Resources and Environmental Management, 1971), 9.
60 J.A. Griffiths, "Drainage Problems in Relation to Manitoba Municipalities," in *Thirteenth Extension Course in Municipal Administration and Public Finance* (Winnipeg: University of Manitoba, Department of Municipal Affairs of Manitoba; Manitoba Municipal Secretary-Treasurers Association, 1960), 26.
61 W.R. Newton, *Resources Management and the Watershed Conservation Districts Act* (Winnipeg: Water Resources Branch, n.d.), 3.
62 Lawrence N. Ogrodnik, *Water Management in the Red River Valley: A History and Policy Review* (Winkler, MB: Lower Red River Valley Water Commission, 1984), 13.
63 Royden Loewen and Gerald Friesen, *Immigrants in Prairie Cities: Ethnic Diversity in Twentieth-Century Canada* (Toronto: University of Toronto Press, 2009), 60.
64 Manitoba, *Whitemud River Watershed Resource Study* (Winnipeg: Department of Mines, Resources and Environmental Management, 1974), 9.
65 Manitoba, *Final Report of the Royal Commission on Local Government Organization and Finance* (Winnipeg: Queen's Printer, 1964), 176-86.
66 MacPherson, *A Brief History of the Whitemud Watershed Committee*, 51; Manitoba, *Whitemud River Watershed Resource Study*, 11.

67  MacPherson, *A Brief History of the Whitemud Watershed Committee*, 39-41, 45; "Watershed Decision Referred to Hutton," *Neepawa Herald*, 27 February 1962, 1.

68  "NADCO to Enquire about ARDA Programme for District," *Neepawa Herald*, 8 January 1963, 1.

69  William Carlyle, "Agricultural Drainage in Manitoba: The Search for Administrative Boundaries," in *River Basin Management: Canadian Experiences*, ed. Bruce Mitchell and James S. Gardner (Waterloo: University of Waterloo, 1983), 279-95; Manitoba Water Commission, *A Review of Agricultural Drainage in Manitoba* (Winnipeg: Department of Mines, Resources and Environmental Management, 1977), 7-8.

70  Donald J. Pisani, *Water and American Government: The Reclamation Bureau, National Water Policy, and the West, 1902-1935* (Berkeley: University of California Press, 2002), 292.

71  Manitoba Water Commission, *A Review of Agricultural Drainage in Manitoba*, 41-43.

72  The Manitoba Water Commission Act, S.M. 1966-67, c. 69; Legislative Assembly of Manitoba, Debates and Proceedings, 27 January 1967, 1st Session, 28th Legislature, 1966-67, 511-17.

73  J.H. Ellis, *The Ministry of Agriculture in Manitoba, 1870-1970* (Winnipeg: Department of Agriculture, 1970), 531-32.

74  Ogrodnik, *Water Management in the Red River Valley*, 14.

75  Manitoba, *Annual Report of Watershed Conservation District(s) of Manitoba* (Winnipeg: Queen's Printer, 1972), 3.

76  Newton, *Resources Management and the Watershed Conservation Districts Act*, 9.

77  Canada, Department of Regional Economic Expansion, *Regional Development Programs* (n.p.: n.p.,1973), 7.

78  Jack D. Giles, *ARDA in Manitoba* (Winnipeg: Department of Agriculture, 1968), 12. The Norquay drain in its original incarnation was among the earliest major drainage projects undertaken by the provincial government. See W.L. Morton, *Manitoba: A History*, 2nd ed. (Toronto: University of Toronto Press, 1967), 206.

79  Giles, *ARDA in Manitoba*, 20.

80  Canada, Department of Regional Economic Expansion, *Regional Development Programs*, 1.

81  Manitoba, Department of Agriculture, and Canada, Department of Regional Economic Expansion, *ARDA III 1972 and 1973: Description and Progress Report* (n.p.: n.p., n.d.), 5.

82  Canada, Agricultural Rehabilitation and Development Administration, *Manitoba ARDA, 1970-1975: Strategy, Area Display, Programs* (Canada: Agricultural Rehabilitation and Development Administration, 1972), 6.

83  LAC, Canada, Department of Regional Economic Expansion Records, RG 124, W 86-87/070, vol. 013, file 740/86027, vol. 2, "Conservation Program Set," *Manitoba Co-Operator*, 10 February 1977, 1.

84  Canada, Agricultural Rehabilitation and Development Administration, *Manitoba ARDA, 1970-1975*, Part 3, "Program Outlines, Resource Programs, Water Conservation Sub-Program," 1-6.

85  W.J. Carlyle, "The Management of Environmental Problems on the Manitoba Escarpment," *Canadian Geographer* 24, 3 (1980): 255-69.

86  Ibid.

87  Canada, Agricultural Rehabilitation and Development Administration, *Manitoba ARDA, 1970-1975*, Part 3, "Program Outlines, Resource Programs, Water Conservation Sub-Program," 4.

88  Manitoba Water Commission, *A Review of Agricultural Drainage in Manitoba*, 56.

89  Ibid., 56-59.

90  Bill Redekop, "Conservation Districts out to Protect the Environment," *Winnipeg Free Press,* 8 January 2006, A5.

91  Ogrodnik, *Water Management in the Red River Valley,* 13-15.

92  Newton, *Resources Management and the Watershed Conservation Districts Act,* 3-4. Although this document is not dated, it is clear from the content that it was prepared after the creation of the Whitemud Watershed Conservation District in 1972 but before the creation of the Turtle Mountain Watershed Conservation District in 1973. A newspaper report containing extensive quotations that correspond to the text in the *Winnipeg Free Press* is dated 14 April 1973.

93  Ibid., 4.

94  Allan Chambers, "A Farmer's Viewpoint of Agricultural Land Drainage," in *Third Annual Western Provincial Conference: Rationalization of Water and Soil Research and Management* (Winnipeg: Manitoba Water Resources Branch, 1985), 149.

95  MacPherson, *A Brief History of the Whitemud Watershed Committee,* 62.

96  Manitoba Water Commission, *A Review of Agricultural Drainage in Manitoba,* 33.

97  Manitoba Conservation Districts Association, "Manitoba Conservation Districts," http:// mcda.ca/.

98  International Institute for Sustainable Development, *Designing Policies in a World of Uncertainty, Change, and Surprise – Adaptive Policymaking for Agriculture and Water Resources in the Face of Climate Change* (Winnipeg: n.p., 2006), 129.

99  Ibid., 132.

100  Loewen and Friesen, *Immigrants in Prairie Cities,* 77.

101  Manitoba, Manitoba Ombudsman, *Report on the Licensing and Enforcement Practices of Manitoba Water Stewardship* (Winnipeg: n.p., 2008).

102  International Joint Commission, *Living with the Red: A Report of the Governments of Canada and the United States on Reducing Flood Impacts in the Red River Basin* (Ottawa: n.p., 2000), 68-72.

## Conclusion

1  AM, Adams Archibald Fonds, MG 12 A1, item 164 A.

2  National Archives, United Kingdom, Kew, Colonial Office, British Columbia, Original Correspondence, CO 60, Russell to Sir John, 2 March 1883. Thank you to John Thistle for drawing this passage to my attention.

3  AM, Manitoba, Drainage Maintenance Boards, Minutes and Office Files, GR 7784, Q 032694, file General, text of an address delivered at the annual meeting of the Union of Municipal Drainage Districts by F.E. Umphrey, Chairman of the Drainage Maintenance Boards, April 1946.

4  E.H. Poyser, "Soil Conservation and the Watershed Approach," in *Eleventh Extension Course in Municipal Administration and Public Finance* (Winnipeg: University of Manitoba, Department of Municipal Affairs of Manitoba; Manitoba Municipal Secretary-Treasurers' Association, 1958), 49-56.

5  Ibid., 55-56.

6  John Warkentin makes a similar point about the importance of long-term memory in relation to arid and semi-arid regions in his article "The Desert Goes North," in *Images of the Plains: The Role of Human Nature in Settlement,* ed. Brian W. Blouet and Merlin P. Lawson (Lincoln: University of Nebraska Press, 1975), 161.

7 Scott St. George and Erik Nielsen, "Paleoflood Records for the Red River, Manitoba, Canada, Derived from Anatomical Tree-Ring Signatures," *The Holocene* 13, 4 (2003): 547-55; Gene Krenz and Jay Leitch, *A River Runs North: Managing an International River* (n.p.: Red River Water Resources Council, 1993), 2. Krenz and Leitch assert that "the Red River Basin always has a water supply problem ... either too much or not enough!"

# Selected Bibliography

ARCHIVAL SOURCES

*Archives of Manitoba (AM)*

Adams Archibald Fonds. AM MG 12 A1.
Church Missionary Society Records. David Anderson, Notes of the Flood at Red River, 1852, by the Bishop of Rupert's Land. AM MG 7 B2 reel A 83.
District of Assiniboia. General Quarterly Court Records. AM MG 2 B4.
Edgar S. Russenholt Fonds. AM boxes P2828-P2841.
H. Albert Hochbaum Fonds. AM boxes P7153-P7195.
John Walter Harris Fonds. AM MG 14 C74.
Lowe Farm Fonds. AM MG 8 A16.
Manitoba. Clerk's Office, Minutes. AM GR 1659.
–. Clerk's Office, Orders in Council. AM GR 589; GR 1530.
–. Department of Natural Resources. Deputy Minister's Files. AM GR 1600.
–. Department of Public Works. Minister's Office Files. AM GR 1607; GR 1609.
–. Deputy Minister's Office Files. AM GR 1610; GR 1611.
–. Dominion Land Surveyors' Notebooks. AM GR 1601.
–. Drainage Maintenance Boards. Minutes and Office Files. AM GR 1617; GR 7784.
–. Executive Council Office. Premier's Office Files. AM GR 553 (John Norquay); GR 1662 (Thomas Greenway); GR 43 (John Bracken, Stuart Garson, Douglas Campbell); GR 45 (Douglas Campbell).
–. Legislative Assembly. Unpublished Sessional Papers. AM GR 174; GR 646.
–. Miscellaneous Land Files. AM GR 7634; GR 7637; GR 7700; GR 7719; GR 7720; GR 7721.
Rural Municipalities of Springfield and Sunnyside Records. AM GR 1670.
Rural Municipality of Westbourne Records. AM GR 1671.
Samuel Taylor Fonds. AM MG 2 C13.
William Pearce Fonds. AM MG 9 A40.

### *Library and Archives Canada (LAC)*

Canada. Privy Council Office Records. LAC RG 2.
–. Department of Finance Records. LAC RG 19.
–. Department of Indian Affairs Records. LAC RG 10.
–. Department of Interior Records. LAC RG 15.
–. Department of Justice Records. LAC RG 13.
–. Department of Public Works Records. LAC RG 11.
–. Department of Regional Economic Expansion Records. LAC RG 124.
–. International Joint Commission Records. LAC RG 51.
–. Water Resources Branch Records. LAC RG 89.

### *The National Archives, United Kingdom*

The National Archives, United Kingdom, Kew, Colonial Office. British Columbia, Original
    Correspondence, CO 60.

### *Saskatchewan Archives Board (SAB)*

George Spence Fonds. SAB, Regina Branch, Coll. 276, Acc. No. 614.

### *University of Alberta Archives (UAA)*

William Pearce Fonds. UAA 74-169.

### OTHER PRIMARY SOURCES

Ducks Unlimited Canada. Manitoba Office Files. Brandon.

### NEWSPAPERS AND PERIODICALS

*Country Guide*
*Emigrant*
*Gladstone Age*
*Manitoba Co-Operator*
*Manitoba Free Press*
*Nor'Wester*
*Western Municipal News*
*Winnipeg Free Press*
*Winnipeg Tribune*

### OTHER SOURCES

Adams, Christopher. *Politics in Manitoba: Parties, Leaders, and Voters.* Winnipeg: University
    of Manitoba Press, 2008.
Amato, Anthony J. "A Wet and Dry Landscape." In *Draining the Great Oasis: An
    Environmental History of Murray County, Minnesota,* ed. Anthony J. Amato, Janet
    Timmerman, and Joseph J. Amato, 1-20. Marshall, MN: Crossings Press, 2001.

Amato, Anthony J., Janet Timmerman, and Joseph A. Amato, eds. *Draining the Great Oasis: An Environmental History of Murray County, Minnesota.* Marshall, MN: Crossings Press, 2001.

Anderson, Roger C. "The Historic Role of Fire in the North American Grassland." In *Fire in North American Tallgrass Prairies,* ed. Scott L. Collins and Linda L. Wallace, 8-18. Norman: University of Oklahoma Press, 1990.

Armstrong, Christopher, Matthew Evenden, and H.V. Nelles. *The River Returns: An Environmental History of the Bow.* Montreal: McGill-Queen's University Press, 2009.

Arthur, W. Brian. *Increasing Returns and Path Dependence in the Economy.* Ann Arbor: University of Michigan Press, 1994.

Artibise, Alan. *Winnipeg: A Social History of Urban Growth, 1874-1914.* Montreal: McGill-Queen's University Press, 1975.

Atwood, C.H. *The Water Resources of Manitoba.* Winnipeg: Manitoba Economic Survey Board, 1938.

Baker, Marilyn. *Symbol in Stone: The Art and Politics of a Public Building.* Winnipeg: Hyperion Press, 1986.

Ball, Norman, ed. *Building Canada: A History of Public Works.* Toronto: University of Toronto Press, 1988.

Bantjes, Rod. *Improved Earth: Prairie Space as Modern Artefact.* Toronto: University of Toronto Press, 2005.

Beyer, Jacquelyn L. "Global Summary of Human Response to Natural Hazards: Floods." In *Natural Hazards: Local, National, Global,* ed. Gilbert White, 265-73. New York: Oxford University Press, 1974.

Blair, Danny. "The Climate of Manitoba." In *The Geography of Manitoba: Its Land and Its People,* ed. John Welsted, John Everitt, and Christoph Stadel, 31-42. Winnipeg: University of Manitoba Press, 1997.

Bloomfield, L.M. *Boundary Waters Problems of Canada and the United States.* Toronto: Carswell, 1958.

Bocking, Stephen. *Nature's Experts: Science, Politics, and the Environment.* New Brunswick, NJ: Rutgers University Press, 2004.

Bogue, Allan G. *From Prairie to Corn Belt: Farming on the Illinois and Iowa Prairies in the Nineteenth Century.* Chicago: University of Chicago Press, 1963.

–. "The Heirs of James C. Malin: A Grassland Historiography." *Great Plains Quarterly* 1, 2 (1981): 105-31.

Bogue, Margaret Beattie. *Patterns from the Sod: Land Use and Tenure in the Grand Prairie, 1850-1900.* Illinois State Historical Library Collections 34. Springfield: Illinois State Historical Society, 1959.

–. "The Swamp Land Act and Wet Land Utilization in Illinois, 1850-1890." *Agricultural History* 25, 4 (1951): 169-80.

Bollens, John C. *Special District Governments in the United States.* Berkeley: University of California Press, 1957.

Boon, T.C.B. "St. Peter's Dynevor, the Original Indian Settlement of Western Canada." *Manitoba Historical Society Transactions* 3 (1952-53): 16-32.

Botts, P.S., and R. Donn. "Temporal Delineation of Wetlands on Gull Point, Presque Isle, Pennsylvania." In *Wetlands: Environmental Gradients, Boundaries, and Buffers,* ed. George Mulamoottil, Barry G. Warner, and Edward A. McBean, 177-91. Boca Raton: Lewis Publishers, 1996.

Bowden, Martyn J. "Desert Wheat Belt, Plains Corn Belt: Environmental Cognition and Behavior of Settlers in the Plains Margin, 1850-1899." In *Images of the Plains: The Role of Human Nature in Settlement,* ed. Brian W. Blouet and Merlin P. Lawson, 189-201. Lincoln: University of Nebraska Press, 1975.

Boyd, Matthew. "Changing Physical and Ecological Landscapes in Southwestern Manitoba in Relation to Folsom (11,000-10,000 BP) and McKean (4,000-3,000 BP) Site Distributions." In *Changing Prairie Landscapes,* ed. Todd A. Radenbaugh and Patrick C. Douaud, 21-38. Regina: Canadian Plains Research Center, 2000.

Brodie, Janine. *The Political Economy of Canadian Regionalism.* Toronto: Harcourt Brace Jovanovich, 1990.

Brooks, Gregory R., et al. *Geoscientific Insights into Red River Flood Hazards in Manitoba: The Final Report of the Red River Flood Project.* Ottawa: Natural Resources Canada, 2003.

Brown, Kate. "Gridded Lives: Why Kazakhstan and Montana Are Nearly the Same Place." *American Historical Review* 106, 1 (2001): 17-48.

Brown, Robert Craig, and Ramsay Cook. *Canada 1896-1921: A Nation Transformed.* Toronto: McClelland and Stewart, 1974.

Bumsted, J.M. *Floods of the Centuries: A History of Flood Disasters in the Red River Valley, 1776-1997.* Winnipeg: Great Plains Publications, 1997.

–. "The Manitoba Royal Commission on Flood Cost Benefit and the Origins of Cost Benefit Analysis in Canada." *American Review of Canadian Studies* 32, 1 (2002): 97-122.

Bumsted, J.R. "Flooding in the Red River Valley of the North." In *Harm's Way: Disasters in Western Canada,* ed. Anthony Rasporich and Max Foran, 239-63. Calgary: University of Calgary Press, 2004.

Burchill, C.S. "The Origins of Canadian Irrigation Law." *Canadian Historical Review* 29, 4 (1948): 353-62.

Burnett, J. Alexander. *A Passion for Wildlife: The History of the Canadian Wildlife Service.* Vancouver: UBC Press, 2003.

Burns, Nancy. *The Formation of American Local Governments: Private Values in Public Institutions.* New York: Oxford University Press, 1994.

Burton, Ian. "Flood-Damage Reduction in Canada." In *Water: Selected Readings,* ed. J.G. Nelson and M.J. Chambers, 77-108. Toronto: Methuen Publications, 1969.

Burton, Ian, Robert W. Kates, and Gilbert F. White. *The Environment as Hazard.* 2nd ed. New York: Guilford Press, 1993.

Campbell, Robert Malcolm. *Grand Illusions: The Politics of the Keynesian Experience in Canada, 1945-1975.* Peterborough: Broadview Press, 1987.

Canada. Department of Resources and Development. *Report on Investigations into Measures for the Reduction of the Flood Hazard in the Greater Winnipeg Area.* Ottawa: Water Resources Division, Red River Basin Investigation, 1953.

–. *Description of the Province of Manitoba.* Ottawa: Department of the Interior, 1893.

–. Dominion Bureau of Statistics. 1921 Census.

–. *Manitoba ARDA, 1970-1975: Strategy, Area Display, Programs.* Agricultural Rehabilitation and Development Administration, 1972.

–. "Peguis First Nation Inquiry Treaty Land Entitlement Claim." In *Indian Claims Commission Proceedings,* vol. 14, 22-30. Ottawa: Indian Claims Commission, 2001.

–. Provincial Secretary's Office. *Papers Relative to the Exploration of the Country between Lake Superior and the Red River Settlement.* London: Her Majesty's Stationery, 1859.

–. *Regional Development Programs.* Department of Regional Economic Expansion, 1973.

–. *Sessional Papers.* Department of the Interior Report, 1896.

Careless, J.M.S. "'Limited Identities' in Canada." *Canadian Historical Review* 50, 1 (1969): 1-10.

Carlson, Anthony E. "'Drain the Swamps for Health and Home': Wetlands Drainage, Land Conservation, and National Water Policy, 1850-1917." PhD diss., University of Oklahoma, 2010.

–. "The Other Kind of Reclamation: Wetlands Drainage and National Water Policy, 1902-1912." *Agricultural History* 84, 4 (2010): 451-78.

Carlyle, William. "Agricultural Drainage in Manitoba: The Search for Administrative Boundaries." In *River Basin Management: Canadian Experiences,* ed. Bruce Mitchell and James S. Gardner, 279-95. Waterloo: University of Waterloo, 1983.

–. "Farm Population in the Canadian Parkland." *Geographical Review* 79, 1 (1989): 13-35.

–. "The Management of Environmental Problems on the Manitoba Escarpment." *Canadian Geographer* 24, 3 (1980): 255-69.

–. "Mennonite Agriculture in Manitoba." *Canadian Ethnic Studies* 13, 2 (1981): 72-97.

–. "The Relationships between Settlement and the Physical Environment in Part of the West Lake Area of Manitoba from 1878 to 1963." MA thesis, University of Manitoba, 1965.

–. "Rural Population Change on the Canadian Prairies." *Great Plains Research* 4, 1 (1994): 65-87.

–. "Water in the Red River Valley of the North." *Geographical Review* 74, 3 (1984): 331-58.

Carroll, John E. *Environmental Diplomacy: An Examination and a Prospective of Canadian-U.S. Transboundary Environmental Relations.* Ann Arbor: University of Michigan Press, 1983.

Carter, Sarah. *Aboriginal People and Colonizers of Western Canada.* Toronto: University of Toronto Press, 1999.

–. "'An Infamous Proposal': Prairie Indian Reserve Land and Soldier Settlement after World War I." *Manitoba History* 37 (1999): 9-21.

–. *Lost Harvests: Prairie Indian Reserve Farmers and Government Policy.* Montreal: McGill-Queen's University Press, 1990.

Carter, V. "Environmental Gradients, Boundaries, and Buffers: An Overview." In *Wetlands: Environmental Gradients, Boundaries, and Buffers,* ed. George Mulamoottil, Barry G. Warner, and Edward A. McBean, 9-17. Boca Raton: Lewis Publishers, 1996.

Castonguay, Stéphane. "Naturalizing Federalism: Insect Outbreaks and the Centralization of Entomological Research in Canada, 1884-1914." *Canadian Historical Review* 85, 1 (2004): 1-35.

Castonguay, Stéphane, and Darin Kinsey. "The Nature of the Liberal Order: State Formation, Conservation, and the Government of Non-Humans in Canada." In *Liberalism and Hegemony: Debating the Canadian Liberal Revolution,* ed. Jean-François Constant and Michel Ducharme, 221-45. Toronto: University of Toronto Press, 2009.

Chacko, J.C. *The International Joint Commission.* New York: Columbia University Press, 1932.

Chambers, Allan. "A Farmer's Viewpoint of Agricultural Land Drainage." In *Third Annual Western Provincial Conference: Rationalization of Water and Soil Research and Management,* 149-59. Winnipeg: Manitoba Water Resources Branch, 1985.

Clark, Andrew Hill. *Acadia: The Geography of Early Nova Scotia to 1760.* Madison: University of Wisconsin Press, 1968.

Clarke, John. "Population and Economic Activity: A Geographical and Historical Analysis." MA thesis, University of Manitoba, 1967.

Collins, Scott L., and Linda L. Wallace, eds. *Fire in North American Tall Grass Prairies.* Norman: University of Oklahoma Press, 1990.

Colpitts, George. *Game in the Garden: A Human History of Wildlife in Western Canada to 1940.* Vancouver: UBC Press, 2002.

Connor, A.J. *The Climate of Manitoba.* Winnipeg: Manitoba Economic Survey Board, 1939.

Constant, Jean-François, and Michel Ducharme, eds. *Liberalism and Hegemony: Debating the Canadian Liberal Revolution.* Toronto: University of Toronto Press, 2009.

Cosgrove, Denis. "The Measures of America." In *Taking Measures across the American Landscape,* ed. James Corner and Alex S. MacLean, 3-14. New Haven: Yale University Press, 1996.

Coupland, R.T. "Mixed Prairie." In *Natural Grasslands: Introduction and Western Hemisphere,* ed. Robert T. Coupland, 151-82. Amsterdam: Elsevier, 1992.

Crerar, Alistair D. "River Basin Management: The Alberta Experience." In *River Basin Management: Canadian Experiences,* ed. Bruce Mitchell and James S. Gardner, 269-78. Waterloo: University of Waterloo, 1983.

Cullingworth, J. Barry. *Urban and Regional Planning in Canada.* New Brunswick, NJ: Transaction Books, 1987.

Cunfer, Geoff, and Dennis Guse. "Ditches." In *Draining the Great Oasis: An Environmental History of Murray County, Minnesota,* ed. Anthony J. Amato, Janet Timmerman, and Joseph A. Amato, 143-58. Marshall, MN: Crossings Press, 2001.

Curtis, Bruce. *The Politics of Population: State Formation, Statistics, and the Census of Canada, 1840-1875.* Toronto: University of Toronto Press, 2001.

Dagenais, Michèle. "The Municipal Territory: A Product of the Liberal Order?" In *Liberalism and Hegemony: Debating the Canadian Liberal Revolution,* ed. Jean-François Constant and Michel Ducharme, 201-20. Montreal: McGill-Queen's University Press, 2009.

Danhoff, Clarence H. *Change in Agriculture: The Northern United States, 1820-1870.* Cambridge, MA: Harvard University Press, 1969.

Danysk, Cecilia. *Hired Hands: Labour and the Development of Prairie Agriculture, 1880-1930.* Toronto: McClelland and Stewart, 1995.

Davidson, Clive B., H.C. Grant, and F. Shefrin. *The Population of Manitoba: A Preliminary Report.* Winnipeg: Manitoba Economic Survey Board, 1938.

Dawson, Carl. *Group Settlement: Ethnic Communities in Western Canada.* Toronto: Macmillan, 1936.

Dawson, Carl, and Eva R. Younge. *Pioneering in the Prairie Provinces: The Social Side of the Settlement Process.* Toronto: Macmillan, 1940.

den Otter, A.A. "Irrigation and Flood Control." In *Building Canada: A History of Public Works,* ed. Norman R. Ball, 143-68. Toronto: University of Toronto Press, 1988.

–. *The Philosophy of Railways: The Transcontinental Railway Idea in British North America.* Toronto: University of Toronto Press, 1997.

Dick, Lyle. *Farmers "Making Good": The Development of Abernethy District, Saskatchewan, 1880-1920.* Calgary: University of Calgary Press, 2008.

Dodds, Gordon B. "The Stream-Flow Controversy: A Conservation Turning Point." *Journal of American History* 56, 1 (1969): 59-69.

Donahue, Brian. *The Great Meadow: Farmers and the Land in Colonial Concord.* New Haven: Yale University Press, 2004.

Donnelly, M.S. *The Government of Manitoba.* Toronto: University of Toronto Press, 1963.

Dorsey, Kurkpatrick. *The Dawn of Conservation Diplomacy: U.S.-Canadian Wildlife Protection Treaties in the Progressive Era.* Seattle: University of Washington Press, 1998.

–. "Scientists, Citizens, and Statesmen: US-Canadian Wildlife Protection Treaties in the Progressive Era." *Diplomatic History* 19, 3 (1995): 407-29.

Doughty, Robin W. *Feather Fashions and Bird Preservation: A Study in Nature Protection.* Berkeley: University of California Press, 1975.

Drache, Hiram. *The Day of the Bonanza: A History of Bonanza Farming in the Red River Valley of the North.* Fargo: North Dakota Institute for Regional Studies, 1964.

Dreisziger, N.F. "The International Joint Commission of the United States and Canada, 1895-1920: A Study in Canadian-American Relations." PhD diss., University of Toronto, 1974.

Dugald Women's Institute History Committee. *Springfield: 1st Rural Municipality in Manitoba, 1873-1973.* Dugald, MB: Dugald Women's Institute, 1974.

Dunlap, Thomas R. *Nature and the English Diaspora: Environment and History in the United States, Canada, Australia, and New Zealand.* Cambridge, UK: Cambridge University Press, 1999.

–. *Saving America's Wildlife: Ecology and the American Mind, 1850-1990.* Princeton: Princeton University Press, 1988.

Easterbrook, W.T. *Farm Credit in Canada.* Toronto: University of Toronto Press, 1938.

Eidse, Lenore, ed. *Furrows in the Valley: Rural Municipality of Morris, 1880-1980.* N.p.: Inter-Collegiate Press, 1980.

Elliot, William. "Artificial Land Drainage in Manitoba: History – Administration – Law." MRM thesis, University of Manitoba, 1977.

Ellis, J.H. *Manitoba Agriculture and Prairie Farm Rehabilitation Activities.* Winnipeg: Department of Agriculture, 1944.

–. *The Ministry of Agriculture in Manitoba, 1870-1970.* Winnipeg: Department of Agriculture, 1970.

–. *The Soils of Manitoba.* Winnipeg: Manitoba Economic Survey Board, 1938.

Ens, Adolf. *Subjects or Citizens? The Mennonite Experience in Canada, 1870-1925.* Ottawa: University of Ottawa Press, 1994.

Ens, Gerhard. *Homeland to Hinterland: The Changing Worlds of the Red River Metis in the Nineteenth Century.* Toronto: University of Toronto Press, 1996.

–. *Volost and Municipality: The Rural Municipality of Rhineland, 1884-1984.* Altona, MB: Friesen Printers, 1984.

Epp-Tiessen, Esther. *Altona: The Story of a Prairie Town.* Altona, MB: D.W. Friesen and Sons, 1982.

Etkin, David, C. Emdad Haque, and Gregory R. Brooks. "Editorial: Toward a Better Understanding of Natural Hazards and Disasters in Canada." *Natural Hazards* 28 (2003): vii-viii.

Evans, Clinton L. *The War on Weeds in the Prairie West: An Environmental History.* Calgary: University of Calgary Press, 2002.

Evenden, Matthew D. *Fish versus Power: An Environmental History of the Fraser River.* New York: Cambridge University Press, 2004.

–. "Precarious Foundations: Irrigation, Environment, and Social Change in the Canadian Pacific Railway's Eastern Section, 1900-1930." *Journal of Historical Geography* 32, 1 (2006): 74-95.

Fafard, Patrick C., and Kathryn Harrison. *Managing the Environmental Union: Intergovernmental Relations and Environmental Policy in Canada.* Montreal: McGill-Queen's University Press, 2000.

Fahrni, Margaret Morton, and W.L. Morton. *Third Crossing.* Winnipeg: Advocate Printers, 1946.

Fallis, Laurence. "The Idea of Progress in the Province of Canada: A Study in the History of Ideas." In *The Shield of Achilles: Aspects of Canada in the Victorian Age,* ed. W.L. Morton, 169-83. Toronto: McClelland and Stewart, 1968.

Farrington, S. Kip. *The Ducks Came Back.* New York: Coward-McCann, 1945.

Ferguson, Barry. *Remaking Liberalism: The Intellectual Legacy of Adam Shortt, O.D. Skelton, W.C. Clark, and W.A. Mackintosh, 1890-1925.* Montreal: McGill-Queen's University Press, 1993.

Fiege, Mark. *Irrigated Eden: The Making of an Agricultural Landscape in the American West.* Seattle: University of Washington Press, 1999.

–. "The Weedy West: Mobile Nature, Boundaries, and Common Space in the Montana Landscape." *Western Historical Quarterly* 35, 1 (2005): 22-47.

Fleming, Sandford. *Reports on Bridging Red River.* Ottawa: n.p., 1879.

Flores, Dan. "Place: An Argument for Bioregional History." *Environmental History Review* 18 (1994): 1-18.

–. "Twenty Years On: Thoughts on Changes in the Land: Indians, Colonists, and the Ecology of New England." *Agricultural History* 78, 4 (2004): 493-96.

Forkey, Neil S. *Shaping the Upper Canadian Frontier: Environment, Society, and Culture in the Trent Valley.* Calgary: University of Calgary Press, 2003.

Foster, Harold D. "Reducing Disaster Losses: The Management of Environmental Hazards." In *Canadian Resource Policies: Problems and Prospects,* ed. Bruce Mitchell and W.R. Derrick Sewell, 209-32. Toronto: Methuen, 1981.

Foster, Janet. *Working for Wildlife: The Beginnings of Preservation in Canada.* Toronto: University of Toronto Press, 1978.

Foster, Kathryn A. *The Political Economy of Special-Purpose Government.* Washington, DC: Georgetown University Press, 1997.

Fowke, Vernon. *The National Policy and the Wheat Economy.* Toronto: University of Toronto Press, 1957.

Francis, R. Douglas. *The Technological Imperative in Canada: An Intellectual History.* Vancouver: UBC Press, 2009.

Frenkel, Stephen. "Old Theories in New Places? Environmental Determinism and Bioregionalism." *Professional Geographer* 46, 3 (1994): 289-95.

Friesen, Gerald. *The Canadian Prairies: A History.* Toronto: University of Toronto Press, 1984.

–. "From 54°40' to Free Trade: Relations between the American Northwest and Western Canada." In *One West, Two Myths: A Comparative Reader,* ed. C.L. Higham and Robert Thacker, 47-63. Calgary: University of Calgary Press, 2004.

–. "Perimeter Vision: Three Notes on the History of Rural Manitoba." In *River Road: Essays on Manitoba and Prairie History,* 197-214. Winnipeg: University of Manitoba Press, 1996.

–. "The Prairie West since 1945: An Historical Survey." In *The Making of the Modern West: Western Canada since 1945,* ed. A.W. Rasporich, 1-35. Calgary: University of Calgary Press, 1984.

–. "The Prairies as Region: The Contemporary Meaning of an Old Idea." In *River Road: Essays on Manitoba and Prairie History,* 165-82. Winnipeg: University of Manitoba Press, 1996.

–. "Space and Region in Canadian History." *Journal of the Canadian Historical Association,* New Series, 16 (2005): 1-22.

Friesen, Gerald (co-written with Jean Friesen). "River Road." In *River Road: Essays on Manitoba and Prairie History,* 3-12. Winnipeg: University of Manitoba Press, 1996.

Friesen, Gerald (co-written with Royden Loewen). "Romantics, Pluralists, Postmodernists: Writing Ethnic History in Prairie Canada." In *River Road: Essays on Manitoba and Prairie History,* 183-96. Winnipeg: University of Manitoba Press, 1996.

Friesen, Jean. "Grant Me Wherewith to Make My Living." In *Aboriginal Resource Use in Canada: Historical and Legal Aspects,* ed. Kerry Abel and Jean Friesen, 141-55. Winnipeg: University of Manitoba Press, 1991.

Friesen, John. "Expansion of Settlement in Manitoba." In *Historical Essays on the Prairie Provinces,* ed. Donald Swainson, 120-30. Toronto: McClelland and Stewart, 1970.

Gates, Paul W. (with Lillian F. Gates). "Canadian and American Land Policy Decisions, 1930." In *The Jeffersonian Dream: Studies in the History of American Land Policy and Development,* ed. Allan G. Bogue and Margaret Beattie Bogue, 148-65. Albuquerque: University of New Mexico Press, 1996.

Giles, Jack D. *ARDA in Manitoba.* Winnipeg: Department of Agriculture, 1968.

Gillis, R.P., and T.R. Roach. *Lost Initiatives: Canada's Forest Industries, Forest Policy, and Forest Conservation.* New York: Greenwood Press, 1986.

Girard, Michel. *L'écologisme retrouvé: Essor et déclin de la Commission de la Conservation du Canada.* Ottawa: Presses de l'Université d'Ottawa, 1994.

Goldsborough, Gordon. *With One Voice: A History of Municipal Governance in Manitoba.* Altona: Friesen Printers, 2008.

Goldsborough, L.G., and G.G.C. Robinson. "Pattern in Wetlands." In *Algal Ecology in Freshwater Benthic Ecosystems,* ed. R.J. Stevenson, M.L. Bothwell, and R.L. Lowe, 77-117. San Diego: Academic Press, 1996.

Goodman, Tristan M. "The Development of Prairie Canada's Water Law, 1870-1940." In *Laws and Societies in the Prairie West, 1670-1940,* ed. Louis A. Knafla and Jonathan Swainger, 266-79. Vancouver: UBC Press, 2005.

Gorman, Hugh S. "Efficiency, Environmental Quality, and Oil Field Brines: The Success and Failure of Pollution Control by Self-Regulation." *Business History Review* 73, 4 (1999): 601-40.

Gosselink, James G., and Edward Maltby. "Wetland Losses and Gains." In *Wetlands: A Threatened Landscape,* ed. Michael Williams, 296-322. Oxford: Basil Blackwell, 1990.

Graham, Robert Michael William. "The Surface Waters of Winnipeg: Rivers, Streams, Ponds, and Wetlands, 1874-1984: The Cyclical History of Urban Land Drainage." MLA practicum, University of Manitoba, 1984.

Greer, Allan, and Ian Radforth. "Introduction." In *Colonial Leviathan: State Formation in Mid-Nineteenth Century Canada,* ed. Allan Greer and Ian Radforth, 3-16. Toronto: University of Toronto Press, 1992.

Griffiths, J.A. "Drainage Problems in Relation to Manitoba Municipalities." In *Thirteenth Extension Course in Municipal Administration and Public Finance,* 23-27. Winnipeg:

University of Manitoba, Department of Municipal Affairs of Manitoba; Manitoba Municipal Secretary-Treasurers Association, 1960.

—. "The History and Organization of Surface Drainage in Manitoba." Paper presented before the Winnipeg Branch, Engineering Institute of Canada, 20 March 1952. Sciences and Engineering Library, University of Manitoba.

Hall, David. *Clifford Sifton.* Vols. 1 and 2. Vancouver: UBC Press, 1981-85.

Hanuta, Irene. "A Reconstruction of Wetland Information in Pre-Settlement Southern Manitoba Using a Geographic Information System." *Canadian Water Resources Journal* 26, 2 (2001): 183-94.

Hard, Herbert A. *Report to the Governor of North Dakota on Flood Control, 1919-1920.* Grand Forks: Normanden Publishing, n.d.

Hargrave, Joseph James. *Red River.* Montreal: John Lovell, 1871.

Harris, Cole. "How Did Colonialism Dispossess? Comments from an Edge of Empire." *Annals of the Association of American Geographers* 94, 1 (2004): 165-82.

—. *The Reluctant Land: Society, Space, and Environment in Canada before Confederation.* Vancouver: UBC Press, 2008.

—. *The Resettlement of British Columbia: Essays on Colonialism and Geographical Change.* Vancouver: UBC Press, 1997.

Harris, Richard Colebrook. *The Seigneurial System in Upper Canada: A Geographical Study.* Montreal: McGill-Queen's University Press, 1984.

Hatvany, Matthew G. *Marshlands: Four Centuries of Environmental Change on the Shores of the St. Lawrence.* Sainte-Foy: Les Presses de l'Université Laval, 2003.

Hays, Samuel P. *Beauty, Health, and Permanence: Environmental Politics in the United States, 1955-1985.* New York: Cambridge University Press, 1987.

—. *Conservation and the Gospel of Efficiency: The Progressive Conservation Movement, 1890-1920.* Cambridge, MA: Harvard University Press, 1959.

Hewes, Leslie. "The Northern Wet Prairie of the United States: Nature, Sources of Information, and Extent." *Annals of the Association of American Geographers* 41, 4 (1951): 307-23.

Hewes, Leslie, and Phillip E. Frandson. "Occupying the Wet Prairie: The Role of Artificial Drainage in Story County, Iowa." *Annals of the Association of American Geographers* 42, 1 (1952): 24-50.

Hewitt, Kenneth, and Ian Burton. *The Hazardousness of a Place: A Regional Ecology of Damaging Events.* Toronto: University of Toronto Press, 1971.

Hildebrand, W., and V. D'Angiolo. "An Assessment of the Alternate Land Use Program for Erosion Control of Steep-Sloped Land along the Riding Mountain Escarpment." In *The Dauphin Papers: Research by Prairie Geographers,* ed. John Welsted and John Everitt, 79-86. Brandon: Department of Geography, Brandon University, 1991.

Hind, Henry Youle. *Reports on the North-West Territory.* Toronto: John Lovell, 1859.

Hochbaum, H. Albert. "Contemporary Drainage within the True Prairie of the Glacial Lake Agassiz Basin." In *Life, Land, and Water: Proceedings from a Conference on the Environmental Studies of the Glacial Lake Agassiz Region,* ed. William Mayer-Oakes, 197-204. Winnipeg: University of Manitoba Press, 1967.

Hollihan, Tony. *Disasters of Western Canada: Courage Amidst the Chaos.* Canada: Folklore Publishing, 2004.

Hughes, Thomas P. *Human-Built World: How to Think about Technology and Culture.* Chicago: University of Chicago Press, 2004.

—. *Networks of Power: Electrification in Western Society, 1880-1930.* Baltimore: Johns Hopkins University Press, 1983.

—. "Technological Momentum." In *Does Technology Drive History? The Dilemma of Technological Determinism,* ed. Merritt Roe Smith and Leo Marx, 101-13. Cambridge, MA: MIT Press, 1994.

Hutchison, Bruce. *The Fraser.* Toronto: Clarke, Irwin, 1950.

International Joint Commission. *Living with the Red: A Report of the Governments of Canada and the United States on Reducing Flood Impacts in the Red River Basin.* Ottawa: n.p., 2000.

Irwin, Robert. "Breaking the Shackles of the Metropolitan Thesis: Prairie History, the Environment, and Layered Identities." *Journal of Canadian Studies* 32, 3 (1997): 98-117.

Isin, Engin F. "The Origins of Canadian Municipal Government." In *Canadian Metropolitics: Governing Our Cities,* ed. James Lightbody, 51-91. Toronto: Copp Clark, 1995.

Jackson, James. *The Centennial History of Manitoba.* Toronto: McClelland and Stewart, 1970.

Johnson, Hildegard Binder. *Order upon the Land: The U.S. Rectangular Survey and the Upper Mississippi Country.* New York: Oxford University Press, 1976.

Johnson, J.K. "'Claims of Equity and Justice': Petitions and Petitioners in Upper Canada, 1815-1840." *Histoire sociale/Social History* 28, 55 (1995): 219-40.

Juliano, K., and S.P. Simonovic. "The Impact of Wetlands on Flood Control in the Red River Valley of Manitoba." Natural Resources Institute, University of Manitoba, 1999.

Kang, B.W. "The Role of Irrigation in State Formation in Ancient Korea." In *A History of Water, Volume 1: Water Control and River Biographies,* ed. Terje Tvedt and Eva Jakobsson, 234-51. London: I.B. Tauris, 2006.

Kaye, Barry. "The Historical Development of the Cultural Landscape of Manitoba to 1870." In *The Geography of Manitoba: Its Land and Its People,* ed. John Welsted, John Everitt, and Christoph Stadel, 79-89. Winnipeg: University of Manitoba Press, 1996.

—. "The Historical Geography of Agriculture and Agricultural Settlement in the Canadian Northwest, 1774-ca. 1830." PhD diss., University of London, 1976.

—. "'The Settlers' Grand Difficulty': Haying in the Economy of the Red River Settlement." *Prairie Forum* 9, 1 (1984): 1-11.

Kaye, B., and D.W. Moodie. "Geographical Perspectives on the Canadian Plains." In *A Region of the Mind: Interpreting the Western Canadian Plains,* ed. Richard Allen, 17-46. Regina: Canadian Plains Research Center, 1973.

Kelsey, Vera. *Red River Runs North!* New York: Harper, 1951.

Kiernan, V.G. "Private Property in History." In *Family and Inheritance: Rural Society in Western Europe,* ed. Jack Goody, Joan Thirsk, and E.P. Thompson, 361-98. Cambridge, UK: Cambridge University Press, 1976.

Kluger, James R. *Turning on Water with a Shovel: The Career of Elwood Mead.* Albuquerque: University of New Mexico Press, 1992.

Krenz, Gene, and Jay Leitch. *A River Runs North: Managing an International River.* N.p.: Red River Water Resources Council, 1993.

Krywulak, Timothy Bruce. "An Archaeology of Keynesianism: The Macro-Political Foundations of the Modern Welfare State in Canada, 1896-1948." PhD diss., Carleton University, 2005.

Kucera, C.L. "Tall Grass Prairie." In *Natural Grasslands: Introduction and Western Hemisphere,* ed. Robert T. Coupland, 227-68. Amsterdam: Elsevier, 1992.

Lahring, Heinjo. *Water and Wetland Plants of the Prairie Provinces.* Regina: Canadian Plains Research Center, 2003.

Lambrecht, Kirk N. *The Administration of Dominion Lands, 1870-1930.* Winnipeg: Hignell Printing, 1991.

Lang, William L. "Bioregionalism and the History of Place." *Oregon Historical Quarterly* 103, 4 (2002): 414-19.

Langston, Nancy. *Forest Dreams, Forest Nightmares: The Paradox of Old Growth in the Inland West.* Seattle: University of Washington Press, 1995.

–. *Where Land and Water Meet: A Western Landscape Transformed.* Seattle: University of Washington Press, 2003.

Latour, Bruno. "Circulating Reference: Sampling the Soil in the Amazon Forest." In *Pandora's Hope: Essays on the Reality of Science Studies,* 24-79. Cambridge, MA: Harvard University Press, 1999.

Laxer, James. *Canada, the U.S., and Dispatches from the 49th Parallel.* Toronto: Doubleday Canada, 2003.

Ledohowski, Edward M. *The Heritage Landscape of the Crow Wing Study Region of Southeastern Manitoba: A Pilot Project.* Winnipeg, MB: Manitoba Culture, Heritage and Tourism, Historic Resources Branch, 2003.

Lee, Lawrence B. "American Influences in the Development of Irrigation in British Columbia." In *The Influence of the United States on Canadian Development: Eleven Case Studies,* ed. Richard A. Preston, 144-63. Durham: Duke University Press, 1972.

–. "The Canadian-American Irrigation Frontier, 1884-1914." *Agricultural History* 40 (1965): 271-83.

Lehr, John. "The Peculiar People: Ukrainian Settlement of Marginal Lands in Southeastern Manitoba." In *Building beyond the Homestead: Rural History on the Prairies,* ed. Ian MacPherson and David Jones, 29-46. Calgary: University of Calgary Press, 1985.

–. "The Process and Pattern of Ukrainian Rural Settlement in Western Canada, 1891-1914." PhD diss., University of Manitoba, 1978.

–. "'Shattered Fragments': Community Formation on the Ukrainian Frontier of Settlement, Stuartburn, Manitoba, 1896-1921." *Prairie Forum* 28, 2 (2003): 219-34.

Lehr, John, and Yossi Katz. "Crown, Corporation, and Church: The Role of Institutions in the Stability of the Pioneer Settlements in the Canadian West, 1870-1914." *Journal of Historical Geography* 21, 4 (1995): 413-29.

–. "Ethnicity, Institutions, and the Cultural Landscapes of the Canadian Prairie West." *Canadian Ethnic Studies* 26, 2 (1994): 70-88.

Leitch, W.G. *Ducks and Men: Forty Years of Co-operation in Conservation.* Winnipeg: Ducks Unlimited Canada, 1978.

Leopold, Aldo. *Round River: From Aldo Leopold's Journal.* New York: Oxford University Press, 1953.

–. *A Sand County Almanac and Sketches Here and There.* New York: Oxford University Press, 1949.

Leopold, Luna B. *A View of the River.* Cambridge, MA: Harvard University Press, 1994.

–. *Water: A Primer.* San Francisco: W.H. Freeman, 1974.

Leopold, Luna B., and Thomas Maddock Jr. *The Flood Control Controversy: Big Dams, Little Dams, and Land Management.* New York: Ronald Press, 1954.

Linton, Jamie. *What Is Water? The History of a Modern Abstraction.* Vancouver: UBC Press, 2010.

Livingstone, David N. *Putting Science in Its Place: Geographies of Scientific Knowledge.* Chicago: University of Chicago Press, 2003.

Loewen, Royden. *Diaspora in the Countryside: Two Mennonite Communities and Mid-Twentieth-Century Rural Disjuncture.* Toronto: University of Toronto Press, 2006.

–. *Ethnic Farm Culture in Western Canada.* Canada's Ethnic Group Series Booklet 29. Ottawa: Canadian Historical Association, 2002.

–. "Ethnic Farmers and the 'Outside' World: Mennonites in Manitoba and Nebraska, 1874-1900." *Journal of the Canadian Historical Association* 1 (1990): 195-213.

–. *Family, Church, and Market: A Mennonite Community in the Old and the New Worlds, 1850-1930.* Toronto: University of Toronto Press, 1993.

–. "On the Margin or in the Lead: Canadian Prairie Historiography." *Agricultural History* 73, 1 (1999): 27-45.

–. "The Quiet on the Land: The Environment in Mennonite Historiography." *Journal of Mennonite Studies* 23 (2005): 151-64.

Loewen, Royden, and Gerald Friesen. *Immigrants in Prairie Cities: Ethnic Diversity in Twentieth-Century Canada.* Toronto: University of Toronto Press, 2009.

Loo, Tina. *Making Law, Order, and Authority in British Columbia, 1821-1871.* Toronto: University of Toronto Press, 1994.

–. "Making a Modern Wilderness: Conserving Wildlife in Twentieth-Century Canada." *Canadian Historical Review* 82, 1 (2001): 92-121.

–. "People in the Way: Modernity, Environment, and Society on the Arrow Lakes." *BC Studies* 142-43 (2004): 161-96.

–. *States of Nature: Conserving Canada's Wildlife in the Twentieth Century.* Vancouver: UBC Press, 2006.

Lowe Farm Chamber of Commerce. *Lowe Farm: 75th Anniversary, 1899-1974.* Altona, MB: D.W. Friesen and Sons, 1974.

Lyons, M.A. *Report and Recommendations on "Foreign Water" and Maintenance Problems in Drainage Maintenance Districts Constituted under the Land Drainage Arrangement Act, 1935, Province of Manitoba.* Winnipeg: King's Printer, 1950.

Macbeth, R.G. *The Selkirk Settlers in Real Life.* Toronto: W. Briggs, 1897.

MacFadyen, Joshua D. "Fashioning Flax: Industry, Region, and Work in North American Fibre and Linseed Oil, 1850-1930." PhD diss., University of Guelph, 2009.

MacPherson, Ian. *The Co-Operative Movement on the Prairies, 1900-1955.* Ottawa: Canadian Historical Association, 1979.

MacPherson, J.C. *A Brief History of the Whitemud Watershed Committee.* Winnipeg: Department of Mines, Resources, and Environmental Management, 1971.

–, ed. *Whitemud River Watershed Conservation District # 1: Watershed Management Seminar Proceedings.* Winnipeg: Department of Mines, Resources, and Environmental Management, 1972.

Mactavish, John C. "The Federal Role in Water Management." Inquiry on Federal Water Policy Research Paper 15, Ottawa, 1985.

Maher, Neil. "'Crazy Quilt Farming on Round Land': The Great Depression, the Soil Conservation Service, and the Politics of Landscape Change on the Great Plains during the New Deal Era." *Western Historical Quarterly* 31, 3 (2000): 319-39.

Malin, James C. *The Grasslands of North America.* Gloucester, MA: Peter Smith, 1967.

–. *History and Ecology: Studies of the Grassland.* Ed. Robert P. Swierenga. Lincoln: University of Nebraska Press, 1984.

Manitoba. *Annual Reports of the Department of Agriculture.* Winnipeg: Queen's Printer.

–. *Annual Reports of the Department of Public Works.* Winnipeg: Queen's Printer.

–. *Annual Reports of the Watershed Conservation District(s) of Manitoba.* Winnipeg: Queen's Printer.

–. Committee on Utilization of Public Lands in the Province of Manitoba. *Report Submitted to the Minister of Agriculture: An Overview.* Winnipeg: Manitoba Department of Agriculture, 1934.

–. Department of Agriculture and Canada Department of Regional Economic Expansion. *ARDA III 1972 and 1973: Description and Progress Report.* N.p: n.p., n.d.

–. *Final Report of the Royal Commission on Local Government Organization and Finance.* Winnipeg: Queen's Printer, 1964.

–. *Journals of the Legislative Assembly.* Winnipeg: Queen's Printer.

–. Land Drainage Arrangement Commission. *Report of the Land Drainage Arrangement Commission Respecting Municipalities Containing Land Subject to Review under "The Land Drainage Act."* Winnipeg: King's Printer, 1936.

–. Manitoba Water Commission. *A Review of Agricultural Drainage in Manitoba.* Winnipeg: Department of Mines, Resources and Environmental Management, 1977.

–. *Report of the Manitoba Drainage Commission.* Winnipeg: King's Printer, 1921.

–. *Summary Report of the Wilson Creek Experimental Watershed Study, 1957-1982.* Winnipeg: Committee on Headwater Flood and Erosion Control, 1983.

–. Water Resources Branch. *Papers of the Third Annual Western Provincial Conference: Rationalization of Water and Soil Research and Management.* Winnipeg: Queen's Printer, 1985.

–. *Whitemud River Watershed Resource Study.* Winnipeg: Department of Mines, Resources and Environmental Management, 1974.

–. *Wilson Creek Experimental Watershed.* Pamphlet. Winnipeg: Province of Manitoba, n.d.

Manitoba Conservation Districts Association. "Manitoba Conservation Districts." http://mcda.ca/.

Manitoba Ombudsman. *Report on the Licensing and Enforcement Practices of Manitoba Water Stewardship.* Winnipeg: n.p., 2008.

Marshall, Daniel P. "An Early Rural Revolt: The Introduction of the Canadian System of Tariffs to British Columbia, 1871-74." In *Beyond the City Limits: Rural History in British Columbia,* ed. Ruth Sandwell, 47-61. Vancouver: UBC Press, 1999.

Matheson, Samuel. "Floods at Red River." In *A Thousand Miles of Prairie: The Manitoba Historical Society and the History of Western Canada,* ed. Jim Blanchard, 239-54. Winnipeg: University of Manitoba Press, 2002.

Matt, Marion. *The Dipper Stick: The History of Drainage in Kent County, Ontario.* Chatham, ON: Thames Arts Centre, 1979.

McCalla, Douglas. "Railways and the Development of Canada West, 1850-1870." In *Colonial Leviathan: State Formation in Mid-Nineteenth-Century Canada,* ed. Alan Greer and Ian Radforth, 192-229. Toronto: University of Toronto Press, 1992.

McCorvie, Mary R., and Christopher L. Lant. "Drainage District Formation and the Loss of Midwestern Wetlands, 1850-1930." *Agricultural History* 67, 4 (1993): 29-39.

McKay, Ian. "The Liberal Order Framework: A Prospectus for a Reconnaissance of Canadian History." *Canadian Historical Review* 81, 4 (2000): 617-45.

McLaughlin, John C. "Progress, Politics, and the Role of Conservation: Wetland Drainage in Ontario." PhD diss., Queen's University, 1995.

McLean, Bill. *Paths to the Living City: The Story of the Toronto and Region Conservation Authority.* Toronto: Toronto and Region Conservation Authority, 2004.

McWilliams, Margaret. *Manitoba Milestones.* Toronto: J.M. Dent and Sons, 1928.

Meindl, Christopher F., Derek H. Alderman, and Peter Waylen. "On the Importance of Environmental Claims-Making: The Role of James O. Wright in Promoting the Drainage of Florida's Everglades in the Early Twentieth Century." *Annals of the Association of American Geographers* 92, 4 (2002): 682-701.

Melosi, Martin. *The Sanitary City: Urban Infrastructure in America from Colonial Times to the Present.* Baltimore: Johns Hopkins University Press, 2000.

Meyer, William. *Americans and Their Weather.* New York: Oxford University Press, 2000.

–. "When Dismal Swamps Became Priceless Wetlands." *American Heritage* 45, 3 (1994): 108-17.

Millard, J. Rodney. *The Master Spirit of the Age: Canadian Engineers and the Politics of Professionalism, 1887-1922.* Toronto: University of Toronto Press, 1988.

Miller, J.R. "Owen Glendower, Hotspur, and Canadian Indian Policy." In *Reflections on Native-Newcomer Relations: Selected Essays,* 107-39. Toronto: University of Toronto Press, 2004.

Milne, Brad. "The Historiography of Métis Land Dispersal, 1870-1890." *Manitoba History* 30 (1995): 30-41.

Mindess, Mervyn. "An Investigation of Soil Moisture Retention in the Red and Assiniboine River Drainage Basins as an Aid to Flood Prevention." MSc thesis, University of Manitoba, 1956.

Mitchell, Bruce, and Dan Shrubsole. *Ontario Conservation Authorities: Myth and Reality.* Waterloo: University of Waterloo Department of Geography Publications Series, 1992.

Mitchner, E.A. "The Development of Western Waters, 1885-1930." Unpublished manuscript, University of Alberta, 1973.

–. "William Pearce and Federal Government Activity in Western Canada, 1882-1904." PhD diss., University of Alberta, 1971.

Mitsch, William, and James Gosselink. *Wetlands,* 3rd ed. New York: John Wiley and Sons, 2000.

Mochoruk, Jim. *Formidable Heritage: Manitoba's North and the Cost of Development 1870 to 1930.* Winnipeg: University of Manitoba Press, 2004.

Moodie, D. Wayne. "Manomin: Historical-Geographical Perspectives on the Ojibwa Production of Wild Rice." *Aboriginal Resource Use in Canada: Historical and Legal Aspects,* ed. Kerry Abel and Jean Friesen, 71-79. Winnipeg: University of Manitoba Press, 1991.

Moodie, D. Wayne, and Barry Kaye. "Indian Agriculture and the Fur Trade Northwest." *Prairie Forum* 11, 2 (1986): 171-83.

Morris, Alexander. *The Treaties of Canada.* Toronto: Belfords, Clarke, 1880.

Morton, A.S. *History of Prairie Settlement.* Toronto: Macmillan, 1938.

Morton, W.L. "Introduction." In *London Correspondence Inward from Eden Colville, 1849-1852,* ed. E.E. Rich, xiii-cxv. London: Hudson's Bay Record Society, 1956.

–. "Introduction." In *Manitoba: The Birth of a Province,* ix-xxx. Winnipeg: Manitoba Record Society Publications, 1984.

–. *Manitoba: A History.* 2nd ed. Toronto: University of Toronto Press, 1967.

–. "Seeing an Unliterary Landscape." In *Contexts of Canada's Past: Selected Essays of W.L. Morton,* ed. A.B. McKillop, 15-25. Toronto: Macmillan, 1980.

Mudry, N., G.H. Mackay, and V.M. Austford. "Flow Control and Flow Regulation Problems on the Assiniboine River." In *River Basin Management: Canadian Experiences*, ed. Bruce Mitchell and James S. Gardner, 297-309. Waterloo: Department of Geography, University of Waterloo, 1983.

Mulamoottil, G., B.G. Warner, and E.A. McBean. "Introduction." In *Wetlands: Environmental Gradients, Boundaries, and Buffers*, ed. George Mulamoottil, Barry G. Warner, and Edward A. McBean, 1-8. Boca Raton: Lewis Publishers, 1996.

Murchie, Robert. *Agricultural Progress on the Prairie Frontier.* Toronto: Macmillan, 1936.

–. *The Unused Lands of Manitoba: Report of Survey.* Winnipeg: Department of Agriculture and Immigration, 1926.

Murray, Stanley Norman. "A History of Agriculture in the Valley of the Red River of the North, 1812 to 1920." PhD diss., University of Wisconsin, 1963.

–.. *The Valley Comes of Age: A History of Agriculture in the Valley of the Red River of the North, 1812-1870.* Fargo: North Dakota Institute for Regional Studies, 1967.

Murton, James. *Creating a Modern Countryside: Liberalism and Land Resettlement in British Columbia.* Vancouver: UBC Press, 2007.

Neufeld, Regina (Doerksen). "Schantzenberg." In *Working Papers of the East Reserve Village Histories, 1874-1910,* ed. John Dyck, 99-107. Steinbach, MB: Hanover Steinbach Historical Society, 1990.

Newson, Malcolm. *Land, Water, and Development: Sustainable Management of River Basin Systems.* 2nd ed. London: Routledge, 1997.

Newton, W.R. *Resources Management and the Watershed Conservation Districts Act.* Winnipeg: Water Resources Branch, n.d.

–. *Watershed Conservation Districts in Manitoba.* N.p.: Watershed Conservation District Boards of Manitoba, n.d.

–. "Watershed Conservation Districts in Manitoba: The Aims, Functions, and Objectives." Paper presented at the 89th Congress of the Engineering Institute of Canada, Winnipeg, 1 October 1975, Conservation and Environment Library, Government of Manitoba.

Nicholas, George P. "Wetlands and Hunter-Gatherer Land Use in North America." In *Hidden Dimensions: The Cultural Significance of Wetland Archaeology,* ed. Kathryn Bernick, 31-46. Vancouver: UBC Press, 1998.

Norrie, Kenneth H. "The National Policy and the Rate of Prairie Settlement: A Review." In *The Prairie West: Historical Readings,* ed. R. Douglas Francis and Howard Palmer, 243-63. Edmonton: University of Alberta Press, 1992.

Nossal, Kim Richard. "Institutionalization and the Pacific Settlement of Interstate Conflict: The Case of Canada and the International Joint Commission." *Journal of Canadian Studies* 18, 4 (1983-84): 75-87.

Nye, David E. *America as Second Creation: Technology and Narratives of New Beginnings.* Cambridge, MA: MIT Press, 2003.

–. "Technology, Nature, and American Origin Stories." *Environmental History* 8, 1 (2003): 8-25.

Ogrodnik, Lawrence N. *Water Management in the Red River Valley: A History and Policy Review.* Winkler, MB: Lower Red River Valley Water Commission, 1984.

Oliver, E.H. *The Canadian North-West: Its Early Development and Legislative Records.* Ottawa: Government Printing Bureau, 1914.

–. "The Institutionalizing of the Prairies." *Transactions of the Royal Society of Canada* 3, 24 (1930): 1-21.

Ommer, Rosemary E., and Nancy J. Turner. "Informal Rural Economies in History." *Labour/Le travail* 53 (2004): 127-58.

O'Neill, Karen M. "Why the TVA Remains Unique: Interest Groups and the Defeat of New Deal River Planning." *Rural Sociology* 67, 2 (2002): 163-82.

Opie, John. *The Law of the Land: Two Hundred Years of American Farmland Policy.* Lincoln: University of Nebraska Press, 1987.

–. "100 Years of Climate Risk Assessment on the High Plains: Which Farm Paradigm Does Irrigation Serve?" *Agricultural History* 63, 2 (1989): 243-69.

O'Riordan, Terence. "Straddling the 'Great Transformation': The Hudson's Bay Company in Edmonton during the Transition from the Commons to Private Property, 1854-1882." *Prairie Forum* 28 (2003): 1-26.

Ostergren, Robert C. "Concepts of Region: A Geographical Perspective." In *Regionalism in the Age of Globalism,* vol. 1, *Concepts of Regionalism,* ed. Lothar Honninghausen, Marc Frey, James Peacock, and Niklaus Steiner, 1-14. Madison: Center for the Study of Upper Midwestern Cultures, 2005.

Owram, Douglas. *Building for Canadians: A History of the Department of Public Works, 1840-1960.* Ottawa: Public Works Canada, 1979.

–. *The Government Generation: Canadian Intellectuals and the State, 1900-1945.* Toronto: University of Toronto Press, 1986.

–. *Promise of Eden: The Canadian Expansionist Movement and the Idea of the West, 1856-1900.* Toronto: University of Toronto Press, 1980.

Paget, A.F., and C.H. Clay. "Multi-Purpose Development of the Fraser River." In *Resources for Tomorrow Conference,* 325-35. Ottawa: Queen's Printer, 1961-62.

Paget, Amelia M. *People of the Plains.* Regina: Canadian Plains Research Center, 2004.

Parsons, James J. "On 'Bioregionalism' and 'Watershed Consciousness.'" *Professional Geographer* 37, 1 (1985): 1-6.

Peers, Laura. *The Ojibwa of Western Canada, 1780 to 1870.* Winnipeg: University of Manitoba Press, 1994.

Percy, David R. "Water Law of the Canadian West: Influences from the Western States." In *Law for the Elephant, Law for the Beaver: Essays in the Legal History of the North American West,* ed. John MacLaren, Hamar Foster, and Chet Orloff, 274-91. Regina: Canadian Plains Research Center, 1992.

–. *Wetlands and the Law in the Prairie Provinces of Canada.* Edmonton: Environmental Law Centre, 1993.

Perruci, Robert, and Joel Gerstl, eds. *The Engineers and the Social System.* New York: John Wiley and Sons, 1969.

Peterson, Thomas. "Manitoba: Ethnic and Class Politics." In *Canadian Provincial Politics: The Party Systems of the Ten Provinces.* 2nd ed., ed. Martin Robin, 61-119. Scarborough: Prentice-Hall, 1978.

Phillips, Alfred Thomas. "Development of Municipal Institutions in Manitoba to 1886." MA thesis, University of Manitoba, 1948.

Piper, Liza. "Colloquial Meteorology." In *Method and Meaning in Canadian Environmental History,* ed. Alan MacEachern and William J. Turkel, 102-23. Toronto: Thompson-Nelson, 2008.

Pisani, Donald. "Beyond the Hundredth Meridian: Nationalizing the History of Water in the United States." *Environmental History* 5, 4 (2000): 466-82.

–. "A Conservation Myth: The Troubled Childhood of the Multiple-Use Idea." *Agricultural History* 76, 2 (2002): 154-71.

–. *To Reclaim a Divided West: Water, Law, and Public Policy, 1848-1902*. Albuquerque: University of New Mexico Press, 1992.

–. *Water and American Government: The Reclamation Bureau, National Water Policy, and the West, 1902-1935*. Berkeley: University of California Press, 2002.

–. *Water, Land, and Law in the West: The Limits of Public Policy, 1850-1920*. Lawrence: University of Kansas Press, 1996.

Platt, Rutherford H. *Disasters and Democracy: The Politics of Extreme Natural Events*. Washington: Island Press, 1999.

Pleva, Edward G. "Multiple Purpose Land and Water Districts in Ontario." In *Comparisons in Resource Management: Six Notable Programs in Other Countries and Their Possible U.S. Application*, ed. Henry Jarrett, 189-207. Baltimore: Johns Hopkins University Press, 1961.

–. "Multiple Purpose Land and Water Districts in Ontario." In *Water: Selected Readings*, ed. J.G. Nelson and M.J. Chambers, 337-51. Toronto: Methuen Publications, 1969.

Potyondi, Barry. *In Palliser's Triangle: Living in the Grasslands, 1850-1930*. Saskatoon: Purich Publishing, 1995.

–. *Selkirk: The First Hundred Years*. Winnipeg: Josten's/National School Services, 1981.

Powell, J. Russell. "River Basin Management in Ontario." In *River Basin Management: Canadian Experiences*, ed. Bruce Mitchell and James S. Gardner, 49-60. Waterloo: University of Waterloo, 1983.

Poyser, E.A. *Your Watershed*. Winnipeg: Department of Agriculture and Immigration, 1958.

Poyser, E.H. "Soil Conservation and the Watershed Approach." In *Eleventh Extension Course in Municipal Administration and Public Finance*, 49-56. Winnipeg: University of Manitoba, Department of Municipal Affairs of Manitoba; Manitoba Municipal Secretary-Treasurers' Association, 1958.

Pratt, Larry. "The State and Province-Building: Alberta's Development Strategy." In *The Canadian State: Political Economy and Political Power*, ed. Leo Panitch, 133-62. Toronto: University of Toronto Press, 1977.

Prince, Hugh. *Wetlands of the American Midwest: A Historical Geography of Changing Attitudes*. Chicago: University of Chicago Press, 1997.

Pritchet, John Perry. *The Red River Valley, 1811-1849: A Regional Study*. New Haven, CT: Yale University Press, 1942.

Pyne, Stephen. *Fire in America: A Cultural History of Wildland and Rural Fire*. Princeton: Princeton University Press, 1982.

Raby, S. "Alberta and the Prairie Provinces Water Board." In *Water: Selected Readings*, ed. J.G. Nelson and M.J. Chambers, 325-35. Toronto: Methuen Publications, 1969.

Rannie, W.F. "Manitoba Flood Protection and Control Strategies." *Proceedings of the North Dakota Academy of Science* 53 (1999): 38-43.

–. "The Red River Flood Control System and Recent Flood Events." *Water Resources Bulletin* 16, 2 (1980): 207-14.

–. *A Survey of Hydroclimate, Flooding, and Runoff in the Red River Basin Prior to 1870*. Open-File Report 3705. Ottawa: Geological Survey of Canada, 1999.

Ray, Arthur. *Indians in the Fur Trade: Their Role as Trappers, Hunters, and Middlemen in the Lands Southwest of Hudson Bay, 1660-1870*. Toronto: University of Toronto Press, 1998.

Ray, Arthur J., Jim Miller, and Frank J. Tough. *Bounty and Benevolence: A History of Saskatchewan Treaties.* Montreal: McGill-Queen's University Press, 2000.

Rees, Ronald. *New and Naked Land: Making the Prairies Home.* Saskatoon: Western Producer Books, 1988.

Regehr, T.D. "Historiography of the Canadian Plains after 1870." In *A Region of the Mind: Interpreting the Western Canadian Plains,* ed. Richard Allen, 87-101. Regina: Canadian Plains Studies Center, 1973.

Reid, John. "Writing about Regions." In *Writing about Canada: A Handbook for Modern Canadian History,* ed. John Schultz, 71-96. Scarborough: Prentice-Hall, 1990.

Reiger, John F. *American Sportsmen and the Origins of Conservation.* New York: Winchester Press, 1975.

Richards, John, and Larry Pratt. *Prairie Capitalism: Power and Influence in the New West.* Toronto: McClelland and Stewart, 1979.

Richardson, Arthur Herbert. *Conservation by the People: The History of the Conservation Movement in Ontario to 1970.* Toronto: University of Toronto Press, 1974.

–. "Ontario's Conservation Authority Program." *Journal of Soil and Water Conservation* 15, 5 (1960): 252-56.

Richtik, James M. "Settlement Process in the 1870s: An Example from Manitoba's Pembina Mountain." *Building beyond the Homestead: Rural History on the Prairies,* ed. Ian MacPherson and David Jones, 7-27. Calgary: University of Calgary Press, 1985.

Riley, R.T. "Memoirs." Photocopied manuscript, Legislative Library of Manitoba, n.d.

Robbins, William G. "Bioregional and Cultural Meaning: The Problem with the Pacific Northwest." *Oregon Historical Quarterly* 103, 4 (2002): 419-28.

Rosen, Christine Meisner, and Joel Tarr. "The Importance of an Urban Perspective in Environmental History." *Journal of Urban History* 20, 3 (1994): 299-310.

Ross, Alexander M., and Terry Crowley. *The College on the Hill: A New History of the Ontario Agricultural College, 1874-1999.* Toronto: Dundurn Press, 1999.

Ross, Eric. *Beyond the River and the Bay.* Toronto: University of Toronto Press, 1970.

Roy, Fernande. *Progrès, harmonie, liberté: Le libéralisme des milieux d'affaires francophones à Montréal au tournant du siècle.* Montréal: Boréal, 1988.

Rudy, Jarett. *Freedom to Smoke: Tobacco Consumption and Identity.* Montreal: McGill-Queen's University Press, 2005.

Russell, Frances. *The Canadian Crucible: Manitoba's Role in Canada's Great Divide.* Winnipeg: Heartland Associates, 2003.

Russenholt, Edgar Stanford. *Heart of the Continent, Being the History of Assiniboia.* Winnipeg: Macfarlane Communications Services, 1968.

Samson, Daniel. *The Spirit of Industry and Improvement: Liberal Government and Rural-Industrial Society, Nova Scotia, 1790-1862.* Montreal: McGill-Queen's University Press, 2008.

Sandwell, Ruth. *Contesting Rural Space: Land Policy and the Practices of Resettlement on Saltspring Island, 1859-1891.* Montreal: McGill-Queen's University Press, 2005.

–. "The Limits of Liberalism: The Liberal Reconnaissance and the History of the Family in Canada." *Canadian Historical Review* 84, 3 (2003): 423-50.

Schiappa, Edward. "Towards a Pragmatic Approach to Definition: 'Wetlands' and the Politics of Meaning." In *Environmental Pragmatism,* ed. Andrew Light and Eric Katz, 209-30. London: Routledge, 1996.

Schultz, S.D. "The Feasibility of Wetland Restoration to Reduce Flooding in the Red River Valley: A Case Study of the Maple (ND) and Wild Rice (MN) Watersheds." Agricultural Economics Report 432, North Dakota State University, Fargo, 1999.

Scott, Geoffrey A.J. "Manitoba's Ecoclimatic Regions." In *The Geography of Manitoba: Its Land and Its People,* ed. John Welsted, John Everitt, and Christoph Stadel, 43-55. Winnipeg: University of Manitoba Press, 1997.

Scott, James. *Seeing like a State: Why Certain Schemes to Improve the Human Condition Have Failed.* New Haven: Yale University Press, 1998.

Sewell, W.R. Derrick. "Changing Approaches to Water Management in the Fraser River Basin." In *Environmental Effects of Complex River Development,* ed. Gilbert F. White, 97-121. Boulder: Westview Press, 1977.

–. "Multi-Purpose Development of Canada's Water Resources." In *Water: Selected Readings,* ed. J.G. Nelson and M.J. Chambers, 261-75. Toronto: Methuen Publications, 1969.

–. "Water Management and Floods in the Fraser River Basin." Department of Geography, University of Chicago, Research Paper 100, 1965.

Shady, Aly M. *Irrigation, Drainage, and Flood Control in Canada.* Ottawa: Irrigation Sector, Canadian International Development Agency, 1989.

Shrubsole, Dan, and Bruce Mitchell. "Practising Sustainable Water Management: Major Themes and Implications." In *Practising Sustainable Water Management: Canadian and International Experiences,* ed. Dan Shrubsole and Bruce Mitchell, 1-25. Cambridge, ON: Canadian Water Resources Association, 1997.

Simpson, Jeffrey. *Spoils of Power: The Politics of Patronage.* Don Mills, ON: Collins Publishers, 1988.

Sinclair, Bruce, Norman R. Ball, and James O. Petersen, eds. *Let Us Be Honest and Modest: Technology and Society in Canadian History.* Toronto: Oxford University Press, 1974.

Smith, Andrew. "Toryism, Classical Liberalism, and Capitalism: The Politics of Taxation and the Struggle for Canadian Confederation." *Canadian Historical Review* 89, 1 (2008): 1-25.

Smith, Goldwin. *Canada and the Canadian Question.* Toronto: Hunter, Rose, 1891.

Smith, Keith D. *Liberalism, Surveillance, and Resistance: Indigenous Communities in Western Canada, 1877-1927.* Edmonton: Athabasca University Press, 2009.

Spencer, Robert, John Kirton, and Kim Richard Nossal, eds. *The International Joint Commission Seventy Years On.* Toronto: T.H. Best Printing, 1981.

Spry, Irene. "The Great Transformation: The Disappearance of the Commons in Western Canada." In *Man and Nature on the Prairies,* ed. Richard Allen, 21-45. Regina: Canadian Plains Research Center, 1976.

St. George, Scott, and Erik Nielsen. "Hydroclimatic Change in Southern Manitoba since A.D. 1409 Inferred from Tree Rings." *Quaternary Research* 58, 2 (2002): 103-11.

–. "Paleoflood Records for the Red River, Manitoba, Canada, Derived from Anatomical Tree-Ring Signatures." *The Holocene* 13, 4 (2003): 547-55.

Steinberg, Theodore. "Do-It-Yourself Deathscape: On the Unnatural History of Disaster in South Florida." *Environmental History* 2, 4 (1997): 414-38.

–. *Slide Mountain.* Berkeley: University of California Press, 1995.

Stetzer, Donald Foster. "Special Districts in Cook County: Toward a Geography of Local Government." Department of Geography, University of Chicago, Research Paper 169, 1975.

Stoesz, Conrad D. "The Post Road." In *Church, Family, and Village: Essays on Mennonite Life on the West Reserve,* ed. Adolf Ens, Jacob E. Peters, and Otto Hamm, 81-90. Winnipeg: Manitoba Mennonite Historical Society, 2001.

Stroud, Ellen. "Does Nature Always Matter? Following Dirt through History." *History and Theory* 42, 4 (2003): 75-81.

Stunden Bower, Shannon. "The Great Transformation? Wetlands and Land Use in Late 19th Century Manitoba." *Journal of the Canadian Historical Association* 15, 1 (2004): 29-47.

–. "Natural and Unnatural Complexities: Flood Control along Manitoba's Assiniboine River." *Journal of Historical Geography* 36, 1 (2010): 57-67.

–. "Watersheds: Conceptualizing Manitoba's Drained Landscape." *Environmental History* 12, 4 (2007): 796-819.

Stutt, Roderick Irwinn. "Water Policy-Making in the Canadian Plains: Historical Factors that Influenced the Work of the Prairie Provinces Water Board (1948-1969)." PhD diss., University of Regina, 1995.

Swyripa, Frances. *Storied Landscapes: Etho-Religious Identity and the Canadian Prairies.* Winnipeg: University of Manitoba Press, 2010.

Tarr, Joel. *The Search for the Ultimate Sink: Urban Pollution in Historical Perspective.* Akron: University of Akron Press, 1996.

Tate, Douglas. "River Basin Development in Canada." In *Canadian Resource Policies: Problems and Prospects,* ed. Bruce Mitchell and W.R. Derrick Sewell, 151-79. Toronto: Methuen, 1981.

Thomas, Craig W. *Bureaucratic Landscapes: Interagency Cooperation and the Preservation of Biodiversity.* Cambridge, MA: MIT Press, 2003.

Thompson, John. "The Bay City Land Dredge and Dredge Works: Perspectives on the Machines of Land Drainage." *Michigan Historical Review* 12, 2 (1986): 21-43.

–. *Wetlands Drainage, River Modification, and Sectoral Conflict in the Lower Illinois Valley, 1890-1930.* Carbondale: Southern Illinois University Press, 2002.

Thompson, John Herd. "Writing about Rural Life and Agriculture." In *Writing about Canada: A Handbook for Modern Canadian History,* ed. John Schultz, 97-119. Scarborough: Prentice-Hall Canada, 1990.

Timmerman, Janet. "Draining the Great Oasis." In *Draining the Great Oasis: An Environmental History of Murray County, Minnesota,* ed. Anthony J. Amato, Janet Timmerman, and Joseph A. Amato, 125-41. Marshall, MN: Crossings Press, 2001.

Tobey, Ronald C. *Saving the Prairie: The Life Cycle of the Founding School of American Plant Ecology, 1895-1955.* Berkeley: University of California Press, 1981.

Treaty 7 Elders and Tribal Council, with Walter Hildebrandt, Sarah Carter, and Dorothy First Rider. *The True Spirit and Original Intent of Treaty 7.* Montreal: McGill-Queen's University Press, 1996.

Tyman, John Langton. *By Section, Township, and Range: Studies in Prairie Settlement.* Brandon: Assiniboine Historical Society, 1972.

–. "Patterns of Western Land Settlement." *Manitoba Historical Society Transactions* 3 (1971-72): 117-35.

–. "Subjective Surveyors: The Appraisal of Farm Lands in Western Canada." In *Images of the Plains: The Role of Human Nature in Settlement,* ed. Brian W. Blouet and Merlin P. Lawson, 75-97. Lincoln: University of Nebraska Press, 1975.

Upham, Warren. *The Glacial Lake Agassiz*. Monographs of the United States Geological Survey 25. Washington, DC: United States Geological Survey, 1895.

Vance, Jonathan F. *Building Canada: People and Projects that Shaped the Nation*. Toronto: Penguin, 2006.

Van Der Goes Ladd, George. *Shall We Gather at the River?* Toronto: United Church of Canada, 1986.

van der Valk, A.G., and C.B. Davis. "The Role of Seed Banks in the Vegetation Dynamics of Prairie Glacial Marshes." *Ecology* 59, 2 (1978): 322-35.

Vennum, Thomas Jr. *Wild Rice and the Ojibway People*. St. Paul: Minnesota Historical Society Press, 1988.

Vileisis, Ann. *Discovering the Unknown Landscape: A History of America's Wetlands*. Washington, DC: Island Press, 1997.

Visvader, Hazel, and Ian Burton. "Natural Hazards and Hazard Policy in Canada and the United States." In *Natural Hazards: Local, National, Global,* ed. Gilbert White, 219-30. New York: Oxford University Press, 1974.

Waggenhoffer, Nickolaus. "Some Socio-Economic Dynamics in South-Eastern Manitoba, with Particular Reference to the Farming Communities within the Local Government Districts of Stuartburn and Piney." MA thesis, University of Manitoba, 1972.

Waisberg, Leo G., and Tim E. Holzkamm. "A Tendency to Discourage Them from Cultivating: Ojibwa Agriculture and Indian Affairs Administration in Northwestern Ontario." *Ethnohistory* 40, 2 (1993): 175-211.

Waiser, Bill. *Saskatchewan: A New History*. Calgary: Fifth House, 2005.

Waiser, William A. "The Government Explorer in Canada, 1870-1914." In *North American Exploration: A Continent Comprehended,* vol. 3, ed. John Logan Allen, 412-60. Lincoln: University of Nebraska Press, 1997.

Walsh, John C., and Steven High. "Rethinking the Concept of Community." *Histoire sociale/Social History* 32, 64 (1999): 255-74.

Warkentin, John. "The Desert Goes North." In *Images of the Plains: The Role of Human Nature in Settlement,* ed. Brian W. Blouet and Merlin P. Lawson, 149-63. Lincoln: University of Nebraska Press, 1975.

–. "Human History of the Glacial Lake Agassiz Region in the 19th Century." In *Life, Land, and Water: Proceedings from a Conference on the Environmental Studies of the Glacial Lake Agassiz Region,* ed. William Mayer-Oakes, 325-37. Winnipeg: University of Manitoba Press, 1967.

–. "Manitoba Settlement Patterns." *Manitoba Historical Society Transactions* 3 (1961): 62-77.

–. "The Mennonite Settlements of Southern Manitoba." PhD diss., University of Toronto, 1960.

–. "Water and Adaptive Strategies in Settling the Canadian West." *Manitoba Historical Society Transactions* 3 (1971-72): 59-73.

Warkentin, John, and Richard Ruggles. *Historical Atlas of Manitoba: A Selection of Facsimile Maps, Plans, and Sketches from 1612-1969*. Winnipeg: Manitoba Historical Society, 1970.

Warren, Louis S. *The Hunter's Game: Poachers and Conservationists in Twentieth-Century America*. New Haven: Yale University Press, 1997.

Watt, W. Edgar. "The National Flood Damage Reduction Program: 1976-1995." *Canadian Water Resources Journal* 20, 4 (1995): 237-48.

Weaver, John C. "Concepts of Economic Improvement and the Social Construction of Property Rights: Highlights from the English-Speaking World." In *Despotic Dominion: Property Rights in British Settler Societies,* ed. John McLaren, A.R. Buck, and Nancy E. Wright, 79-102. Vancouver: UBC Press, 2005.

–. *The Great Land Rush and the Making of the Modern World, 1650-1900.* Montreal/Kingston: McGill-Queen's University Press, 2003.

Weaver, J.E. *North American Prairie.* Lincoln: Johnson Publishing, 1954.

Webb, Walter Prescott. *The Great Plains.* Boston: Ginn, 1959.

Weber, Thomas E. "On Being Downstream from Everyone Else." *Canadian Water Resources Journal* 4, 3 (1979): 75-81.

Welsted, John, John Everitt, and Christoph Stadel. *The Geography of Manitoba: Its Land and Its People.* Winnipeg: University of Manitoba Press, 1997.

Wentz, Richard, and Nicoletta Barrie, eds. *Return to Big Grass.* N.p.: Ducks Unlimited, 1986.

Wescoat, James L. Jr. "Integrated Water Development: Water Use and Conservation Practice in Western Colorado." Department of Geography, University of Chicago, Research Paper 20, 1984.

–. "'Watersheds' in Regional Planning." In *The American Planning Tradition: Culture and Policy,* ed. Robert Fishman, 147-71. Washington, DC: Woodrow Wilson Center Press, 2000.

Wescoat, James L. Jr., and Gilbert F. White. *Water for Life: Water Management and Environmental Policy.* Cambridge, UK: Cambridge University Press, 2003.

West, Elliot. "Against the Grain: State-Making, Cultures, and Geography in the American West." In *One West, Two Myths: A Comparative Reader,* ed. C.L. Higham and Robert Thacker, 1-21. Calgary: University of Calgary Press, 2004.

White, Gilbert F. "Contributions of Geographical Analysis to River Basin Development." In *Readings in Resource Management and Conservation,* ed. Ian Burton and Robert W. Kates, 375-94. Chicago: University of Chicago Press, 1965.

–. "Natural Hazards Research: Concepts, Methods, and Policy Implications." In *Natural Hazards: Local, National, Global,* ed. Gilbert White, 3-16. New York: Oxford University Press, 1974.

–. "Role of Geography in Water Resources Management." In *Man and Water: The Social Sciences in Management of Water Resources,* ed. L. Douglas James, 102-21. Lexington: University of Kentucky Press, 1974.

–. *Strategies of American Water Management.* Ann Arbor: University of Michigan Press, 1969.

–. "Watershed and Streams of Thought." In *Reviews in Ecology: Desert Conservation and Development,* ed. H.N. Barakat and A.K. Hegazy, 89-98. Cairo: Metropole, 1997.

White, Richard. "The Nationalization of Nature." *American Historical Review* 86, 3 (1999): 976-86.

–. *The Organic Machine.* New York: Hill and Wang, 1995.

–. "Tempered Dreams." In *Inventing for the Environment,* ed. Arthur Molella and Joyce Bedi, 3-10. Cambridge, MA: MIT Press, 2003.

Whitney, Gordon G., and Joseph P. Decant. "Government Land Office Surveys and Other Early Land Surveys." In *The Historical Ecology Handbook: A Restorationist's Guide to Reference Ecosystems,* ed. Dave Egan and Evelyn A. Howell, 147-72. Washington, DC: Island Press, 2005.

Wiebe, Robert H. *The Search for Order, 1877-1920.* New York: Hill and Wang, 1967.

Wiken, E.B. *Terrestrial Ecozones of Canada.* Ecological Land Classification Series 19. Hull, QC: Environment Canada, 1986.

Williams, Michael. "Agricultural Impacts in Temperate Wetlands." In *Wetlands: A Threatened Landscape,* ed. Michael Williams, 181-216. Oxford: Basil Blackwell, 1990.

–. "The End of Modern History?" *Geographical Review* 88, 2 (1998): 275-300.

–. "Protection and Retrospection." In *Wetlands: A Threatened Landscape,* ed. Michael Williams, 323-53. Oxford: Basil Blackwell, 1990.

–. "Understanding Wetlands." In *Wetlands: A Threatened Landscape,* ed. Michael Williams, 1-41. Oxford: Basil Blackwell, 1990.

Willoughby, William R. "Expectations and Experience, 1909-1979." In *The International Joint Commission Seventy Years On,* ed. Robert Spencer, John Kirton, and Kim Richard Nossal, 24-42. Toronto: T.H. Best Printing, 1981.

Wilson, Catherine Anne. *Tenants in Time: Family Strategies, Land, and Liberalism in Upper Canada, 1799-1871.* Montreal: McGill-Queen's University Press, 2009.

Wilson, Robert M. *Seeking Refuge: Birds and Landscapes of the Pacific Flyway.* Seattle: University of Washington Press, 2010.

–. "Seeking Refuge: Making Space for Migratory Waterfowl and Wetlands along the Pacific Flyway." PhD diss., University of British Columbia, 2003.

Wiseman, Nelson. "The Pattern of Prairie Politics." In *The Prairie West: Historical Readings.* 2nd ed., ed. R. Douglas Francis and Howard Palmer, 640-60. Edmonton: University of Alberta Press, 1992.

Wittfogel, Karl. *Oriental Despotism: A Comparative Study of Total Power.* New Haven: Yale University Press, 1957.

Wohl, Ellen E. *Disconnected Rivers: Linking Rivers to Landscapes.* New Haven: Yale University Press, 2004.

Wood, J. David. *Making Ontario: Agricultural Colonization and Landscape Colonization before the Railway.* Montreal: McGill-Queen's University Press, 2000.

–. *Places of Last Resort: The Expansion of the Farm Frontier into the Boreal Forest in Canada, c. 1910-1940.* Montreal: McGill-Queen's University Press, 2007.

Worster, Donald. "Climate and History: Lessons from the Great Plains." In *Earth, Air, Fire, Water: Humanistic Studies of the Environment,* ed. Jill Ker Conway, Kenneth Keniston, and Leo Marx, 51-77. Amherst: University of Massachusetts Press, 1999.

–. *Dust Bowl: The Southern Plains in the 1930s.* New York: Oxford University Press, 1979.

–. "History as Natural History: An Essay on Theory and Method." *Pacific Historical Review* 53 (1984): 1-19.

–. *Nature's Economy: The Roots of Ecology.* San Francisco: Sierra Club Books, 1977.

–. *Rivers of Empire: Water, Aridity, and the Growth of the American West.* New York: Pantheon Books, 1985.

–. "The Second Colorado River Expedition: John Wesley Powell, Mormonism, and the Environment." In *Surveying the Record: North American Scientific Exploration to 1930,* ed. Edward C. Carter II, 317-28. Philadelphia: American Philosophical Society, 1999.

–. "Two Faces West: The Development Myth in Canada and the United States." In *Terra Pacifica: People and Place in the Northwest States and Western Canada,* ed. Paul W. Hirt, 71-91. Pullman: Washington State University Press, 1998.

Wynn, Graeme. "Forging a Canadian Nation." In *North America: The Historical Geography of a Changing Continent,* ed. Robert D. Mitchell and Paul A. Groves, 373-409. Totowa, NJ: Rowman and Littlefield, 1987.

—. "'Shall We Linger along Ambitionless?' Environmental Perspectives in British Columbia." *BC Studies* 142-43 (2004): 5-67.

Zeller, Suzanne. *Inventing Canada: Early Victorian Science and the Idea of a Transcontinental Nation.* Toronto: University of Toronto Press, 1987.

# Index

*Note:* "(f)" after a page number indicates a figure; "(t)" after a page number indicates a table

Aboriginal peoples: 15-16, 20-21, 43-49; Assiniboine, 44; Cree, 44; Ojibwa, 44; relations with the province, 45; treaties, 44-45

Agricultural Rehabilitation and Development Act (1961). *See* Agricultural and Rural Development Act (ARDA)

Agricultural and Rural Development Act (ARDA), 158-59

agricultural settlement, 21, 24, 48, 164

Alonsa Conservation District, 160

Alternative Land Use Program. *See* Agricultural and Rural Development Act (ARDA)

American Homestead Act (1862), 21, 23

Archibald, Adams (Lieutenant Governor), 164

Assiniboine Delta, 6

Assiniboine River, 1, 6, 7, 8, 22, 47, 68, 82, 90, 117, 118, 152, 169

Better Terms Agreement (1885), 56-57, 60

Big Grass Marsh, 27, 54, 123(f), 137-38, 152, 199*n*46; drainage of, 110, 121-22; restoration, 111, 121, 137; and waterfowl, 123-32, 126(f), 129(f)

bioregionalism, 14, 15, 110-38

bogs. *See* wetlands, bogs

boreal forest, 4, 6

Boundary Waters Treaty (1909), 111, 115, 174. *See also* International Joint Commission

Bowman, H.A., 23, 25, 95, 115

Bracken, John, 86, 101

British, 15, 35, 41, 101

British North America Act, 16, 22, 45, 65, 68, 116, 174

Brown, C.P., 83

Cameron, K.M., 120

Canadian Shield. *See* Precambrian Shield

Clubb, W.R., 120

Committee on the Utilization of Public Land in the Province of Manitoba, 124, 200*n*66

Comprehensive Soil and Water Conservation Program. *See* Agricultural and Rural Development Act (ARDA)

Confederation, 16, 22, 45, 55, 56, 61, 141, 143

conservation district system, 158-62, 159(f)

Conservation Districts Act (1976), 161, 174
corruption, 28, 31, 53, 77, 84. *See also* patronage

**D**epartment of Agriculture (Manitoba) 140, 144, 157, 158, 170
Department of Agriculture (US), 85, 114
Department of Highways, 157
Department of the Interior (Canada), 35, 57, 66, 68, 69, 71, 113, 116, 118, 133; Reclamation Service, 68; Water Powers Branch, 66, 118, 160, 162
Department of Mines and Natural Resources, 125, 131, 134, 157
Department of Provincial Lands, 58
Department of Public Works: and American drainage, 115, 120; complaints to, 29; design of drainage districts, 76, 78-83; financial assistance from, 31; Reclamation Branch, 70, 74, 81(f), 95; responsibility for drainage, 85, 134, 140, 157; views on drainage infrastructure, 22-26, 95, 103, 105-6, 112, 137
Department of Regional Economic Expansion (Canada), 158
district systems, 2, 148
Dominion Land Survey, 23, 32, 33, 53-54, 71-72, 80(f)
Dominion Lands Act (1872), 21, 23, 29, 37, 49, 96, 174
Dominion Lands Board, 64
dominion-provincial relations. *See* jurisdiction, provincial-federal
drainage: agricultural, 26, 29, 32, 48, 99, 141, 142-43, 159; federal funding of, 56, 157-58, 163; government-assisted, 9, 12, 33; municipal funding of, 32, 145-46; provincial funding of, 39, 99, 105-8, 112, 139, 149-51, 163, 172(t), 180*n*32, 195*n*79; recreational uses, 30(f); road, 23, 29, 98-99, 182*n*25; surface ditches, xiv, 28-29, 30(f), 33-39, 35(f), 46, 58, 68, 90, 99, 101, 119, 122, 131, 139, 143-48, 182*n*25; urban, 117-18, 163, 198*n*31
Drainage Act (1880), 26-27, 28, 53
drainage districts, establishment of, 9, 15, 31-43, 34(f), 40(f), 47-49, 68-69, 76-79,

79(f), 81(f), 82, 92(f), 93(f), 113, 123, 173(t), 183*n*40, 191*n*4; plans, xix, 81(f), 82, 132-33; public opinion about, 76, 84-85, 123; and watersheds, 87-90, 88(f), 94, 102, 133. *See also* Sullivan Commission
drainage lands management, 53-56
drainage maintenance boards, 148, 150
drains: construction and design of, 20, 47, 76, 77-78, 149-50, 182*n*25; digging of, 27(f); double dyke, 101-5, 102(f), 117, 133; maintenance, 143-45, 148-49, 150
dredges, xii, 34, 43(f)
drought, 3, 121, 124-25, 130, 131, 137, 153, 169
Duck Factory No. 1, 111, 136
Ducks Unlimited, 14, 17, 111-12, 121, 127-38, 129(f), 152, 167, 174, 191*n*11, 198*n*40
Ducks Unlimited (Canada), 127-29, 174
Dutch, 41, 41(t), 185*n*70. *See also* Mennonites
dykes, xii, 8, 101-5, 117, 133, 142, 160

**e**cological commons, 13-16, 20, 36, 38, 39, 42-43, 47, 49, 97, 104, 107, 138
Elliot, Charles Gleason, 85, 87, 89, 100, 101, 192*n*21
engineering, 32, 76-77; engineers as experts, 78-79, 79(f), 82, 84-85, 86-87, 95, 139
English, 41, 41(t)
ethnic communities, 12, 15, 36, 41(f), 42-43, 49
erosion. *See* soils

**F**inlayson, John N., 105
Finlayson Royal Commission, 105, 148-50, 174
flooding: Assiniboine, 68; Big Grass Marsh, 122; Drainage District No. 1, Red River Settlement, 82-83; Drainage District No. 2, Manitoba Escarpment, 90-91; drainage districts to combat, 78; and highland-lowland dispute, 115; impacts of, xix, 12, 15, 25-26, 38; Red River, 103, 141-42, 163; Roseau River, 111, 116-17

French, 15, 39, 41, 41(t)

Germans, 39, 41, 41(t), 185*n*70. *See also*
     Mennonites
government assistance: to Aboriginal
     peoples, 45-48; federal, 121-22, 157;
     municipal, 149; provincial, 9, 11, 30-31,
     38, 42, 45-48, 76, 165, 180*n*32
Grain Growers Association of Manitoba,
     97
Greenway, Thomas, 28, 83
Grills, H., 85. *See also* Sullivan Commission

Harrison, David, 83
highlander: conflicts with lowlanders over
     drainage, xvi, 76-109, 114-15, 119-20, 137,
     140, 144, 153-54, 160, 166-67, 192*n*35;
     farms, 118; opinions about watershed
     management, 18, 94-98, 103, 106, 108,
     140, 160-61, 167
Hind, Henry Youle, 19, 49
Holland, John, 105. *See also* Finlayson
     Royal Commission
homesteaders, 21, 23, 25, 33, 37-38, 54, 59,
     70, 82, 96, 100, 104, 121, 165
hunting, xii, 21, 24, 44, 45, 125, 127, 130

Immigration, 1, 22, 24, 28-29, 35, 60,
     181*n*15
Ingram, Alex, 1-2, 4
International Joint Commission, 110-38,
     152, 163,
Irish, 41, 41(t)
irrigation, 9, 33, 52, 64-68, 74-75, 94, 116,
     152, 174

jurisdiction: Canada-US, 14, 111-12, 115-
     18, 137, 152; municipal-provincial, 14,
     139, 146-47, 156, 160-61; provincial
     government departments, 133-34, 140;
     provincial-federal, 14, 16, 19-20, 22, 45-
     49, 56-61, 62(f), 63(f), 67-75, 116-17,
     133-34, 138, 142-43, 157-58, 165, 168; US
     state-federal, 52-53; water in relation to,
     13-14, 38

Lake Agassiz, 4, 88, 121, 199*n*58

Lake Manitoba, xii
Lake Winnipeg, 2, 4, 44, 46
Land Drainage Act (1895): and creation
     of drainage districts, 33, 39, 78-79, 84,
     107-8, 113, 122, 124, 133, 137, 145-46; and
     Department of Public Works, 83; and
     engineering expertise, 76-77; results of,
     105, 139-40, 145-46, 156. *See also* drain-
     age districts
Land Drainage Arrangement Act (1935),
     104, 174
Laurentian Shield. *See* Precambrian
     Shield
Legislative Committee on Drainage, 86,
     94-98, 103, 108, 118
Leopold, Aldo, xi-xiii, 130-31
levees, 101, 102, 104
lowlanders: conflicts with highlanders
     over drainage, xvi, 76-109, 114-15, 119-
     20, 137, 140, 144, 153-54, 160, 166-67,
     192*n*35; opinions about watershed
     management, 90-91, 94, 96-98, 102-3,
     106, 108
Lyons, M.A., 24, 108, 133, 150-53, 174,
     195*n*74, 204*n*44; inquiry into drainage
     system, 103-6

Macdonald, Hugh John, 83
Magrath, Charles A., 119
Manitoba: Act (1870), 50, 174; territory,
     65, 65(f)
Manitoba Drainage Commission (1919).
     *See* Sullivan Commission
Manitoba Drainage Company, 83, 122
Manitoba Escarpment, 2, 4, 6, 35, 87,
     90-91, 153, 182
Manitoba Executive Council, 27, 40
Manitoba Lowlands: effects of highlands
     drainage on, 91, 95-96, 99, 101, 104-6,
     144, 153; flooding on, 90-91; physical
     features, 4, 5(f), 6, 88, 90, 194*n*52. *See*
     *also* lowlanders
Manitoba Royal Commission on Local
     Government Organization and Finance,
     156
Manitoba Water Commission (1967), 140,
     156-57, 159-61, 174

Manitoba Water Resources Branch, 154, 160

marshes. *See* wetlands, marshes

marshes, drainage of. *See* wetlands, drainage of

McDiarmid, J.S., 128, 131

Mennonite West Reserve, 38, 40(f)

Mennonites, 15, 20, 36-44, 40(f), 41(t), 49, 185*n*70; group settlement, 37; land management system, 37

Métis, 21, 51

Migratory Bird Treaty, 125, 174

Montcalm, Rural Municipality of, 39, 40(f), 41(t)

More Game Birds in America Foundation, 127

Morris, Rural Municipality of, 40(f), 41(t), 76, 146, 193*n*43

Municipal Act, The, 145

municipalities, 40(f), 41(t), 45, 47, 76, 84, 92(f), 94-96, 101, 105, 113, 116, 122, 130, 137, 141, 145-50, 154-57, 160-62, 167, 168, 172(t), 174, 185*n*70, 193*n*43, 193*n*44, 204*n*32; boundaries, 12, 20, 31, 33, 38, 39, 42, 48; conflicts between, 14; drain maintenance by, 147-48; governments, 31-43, 48, 98-100, 139, 145, 170; incorporation, 31; patronage, 28; representation on drainage maintenance district boards, 148; taxation of residents, 146-47, 155

muskeg, 23, 44, 164

Newton, William R., 139, 154, 160

Norquay, John, 56-57, 83, 143-44, 149, 181*n*15

Norquay Floodway, 158, 206*n*78

Norris, T.C., 84, 85, 86

North Norfolk, Rural Municipality of, 47, 193*n*44

North-West Territories, 65, 65(f)

Northwest Irrigation Act (1894), 64-68, 74, 116, 174

Ochre River, Rural Municipality of, 96, 193*n*44

patronage, xv, 16, 28, 77, 83-87, 108, 166

Pearce, William, 64-66

Portage la Prairie, Rural Municipality of, 47, 146

Precambrian Shield, 4, 35-36, 87

precipitation, 2, 3, 6-7, 7(t), 25, 51, 69, 90, 103, 130, 150, 151, 153, 169

private property ideal, 9, 12-13, 32, 77, 90, 96-98, 138, 167-68

railways, 22, 66, 71, 141-42, 165, 182*n*25

reclamation, 53, 56, 61, 68, 85, 95-96, 121. *See also* Department of Public Works, Reclamation Branch

Red River: drainage along, 68, 90, 115, 117, 142, 158; physical features, 6, 7, 44, 87, 110, 112, 121, 145; valley, 4, 36. *See also* flooding, Red River

Red River Floodway, 142, 152, 159(f), 163

Red River Settlement, 51, 82, 125, 143

Red River Valley Drainage and Improvement Association, 40-42, 41(t), 84, 94, 154, 185*n*69

reserves: Aboriginal, 15, 20, 44, 45-49; Mennonite, 36-38, 40(f), 44, 49

Resource Conservation Districts Act (1970), 157, 159, 161, 174

Rhineland, Rural Municipality of, 38-40, 40(f), 41(t), 45, 193*n*43

Riding Mountain-Whitemud River Watershed Committee, 154, 155

roads: allowances, 23, 24, 32; construction, 12, 31-32, 182*n*25; system, 22, 98-99. *See also* drainage, road

Roblin, R.P., 39, 61, 83, 87

Rocky Mountains, 2, 35, 165

Roland, Rural Municipality of, 41(t), 94, 146, 193*n*43

Roseau River, 17, 110-21, 114(f), 130, 133, 137-38, 152, 167, 197*n*18

Russell, Lindsay (Dominion Land Surveyor), 53-54, 164-65

Russian, 41, 41(t)

science, influence of, 2, 14, 16, 73, 82, 86, 95, 96, 101, 127

Scottish, 41, 41(t)

Soil Conservation Sub-Program. *See* Agricultural and Rural Development Act (ARDA)

soils: conservation, 156, 159; erosion, 17-18, 140, 144, 148, 153-54, 167; permeability, 2, 6, 25, 169; quality of, 1, 6, 7-8, 36, 51, 66, 111, 113, 124, 143-45; sedimentation, 90, 144-45, 153

Souris River, 114(f), 118-19, 121

Southwest Uplands, 4, 6, 35

Spalding, John, 105. *See also* Finlayson Royal Commission

Stanley, Rural Municipality of, 7, 95, 193*n*44

Sullivan, J.G., 85, 108, 115, 117, 153, 193*n*44, 204*n*32. *See also* Sullivan Commission

Sullivan Commission (1919), 72, 85, 86, 91, 94-98, 101, 102, 104, 115-16, 120, 147, 174, 193*n*44

Swamp Lands Act (Canada) (1885), 56, 58, 60, 82, 174

Swamp Lands Act (United States) (1849), 53

swamplands. *See* wetlands, swamps

Swamplands commissioners, 56-59, 71

Territorial Grain Growers Association. *See* Grain Growers Association of Manitoba

Thompson, J.A., 85. *See also* Sullivan Commission

townships, 21-22, 39, 54, 81(f), 82, 91; plans, 79, 80(f), 81-82; survey, 21, 23, 33, 50, 139, 166

transboundary flooding, 17, 38, 116-18, 167

transboundary water management, 14, 17, 111, 115-20, 133, 137-38. *See also* Boundary Waters Treaty

transportation, 3, 11, 22, 24, 26, 75, 105, 155. *See also* railways; roads

treaties, 111, 115, 125, 174; Aboriginal, 44-45

Ukrainians, 15, 113

Umphrey, F.E. (Drainage Maintenance Boards Chair), 23, 136, 148, 149, 150-51, 167, 182*n*21

Union of Municipal Drainage Maintenance Districts, 149, 150, 152, 154

US Department of Agriculture, 85, 114

Victoria, Rural Municipality of, 7, 95, 193*n*44

Water Conservation Sub-Program. *See* Agricultural and Rural Development Act (ARDA)

waterfowl, 17, 21, 111, 124-38, 167, 201*n*111. *See also* Ducks Unlimited

Watershed Conservation Districts Act (1959), 154, 156, 157, 159, 161, 174

watershed management: Canada-US issues, 114, 116-18, 120; debates about, 17-18, 77, 87, 91, 94-98, 106-8, 153-63, 167-68; infrastructure, 64, 100, 102-4, 133, 136-38; and soil erosion, 152-62

Watershed and Soil Conservation Authorities Act (1958), 154, 156, 174

Westbourne, Rural Municipality of, 99, 105, 130, 154; Westbourne Municipal Council, 122

wetlands, 188*n*24, 188*n*39, 190*n*79, 191*n*11, 198*n*31; bogs, xi, xii, xiii, 4, 19, 49; definition and identification of, 3-4, 51, 54, 73; drainage of, xvi, 26-27, 51, 54, 67-71, 78, 111, 121-22, 135, 137, 198*n*31; ecology of, 7, 21, 50-51; jurisdiction over, xv, 50-75, 62(f); marshes, 26-27, 54, 90, 111; restoration of, 112, 124-27, 130-38, 175*n*5, 191*n*11; swamps, 4, 50, 56-62, 62(f), 67, 70-74, 78-79, 82-83, 116, 122, 164, 174, 188*n*24, 188*n*39; utility of, 20-21, 25, 44, 50, 51, 71, 111, 121, 124, 132, 136

White Mud River, 121

Whitemud Watershed Conservation District, 157

wildlife, xii, 128, 130-31, 134-35, 157. *See also* waterfowl